CHALCOGEN CHEMISTRY

Edited by **Peter Papoh Ndibewu**

Chalcogen Chemistry
http://dx.doi.org/10.5772/intechopen.73066
Edited by Peter Papoh Ndibewu

Contributors

Soufiane Zerraf, Said Belaaouad, Mustafa Belhabra, Kheireddine Aziz, Malika Tridane, Hicham Moutaabbid, Mohammed Moutaabbid, Dan Xu, Thies Thiemann, Mohamed Loukil, Surjani Wonorahardjo, I Wayan Dasna, Fariati Fariati, Peter Papoh Ndibewu

First published in London, United Kingdom, 2019 by IntechOpen
IntechOpen is the global imprint of INTECHOPEN LIMITED, registered in England and Wales, registration number: 11086078, The Shard, 25th floor, 32 London Bridge Street
London, SE19SG – United Kingdom
Printed in Croatia

British Library Cataloguing-in-Publication Data
A catalogue record for this book is available from the British Library

Additional hard copies can be obtained from orders@intechopen.com

Chalcogen Chemistry, Edited by Peter Papoh Ndibewu
p. cm.
Print ISBN 978-1-78985-019-2
Online ISBN 978-1-78985-020-8

For EU product safety concerns:
IN TECH d.o.o., Prolaz Marije Krucifikse Kozulić 3, 51000 Rijeka, Croatia,
info@intechopen.com or visit our website at intechopen.com.

Meet the editor

Prof Peter Papoh Ndibewu holds a PhD in Chemistry (NMMU Port Elizabeth, South Africa) and is a former visiting scholar at the University of Saskatchewan (Saskatoon, Canada). He currently teaches analytical, physical, and metallurgical chemistry at the Tshwane University of Technology, Pretoria, South Africa. He is an acclaimed researcher and has published widely on multidisciplinary themes of interest. His current research activities focus on spectroscopy and composite materials characterization, and soil hydrogeochemistry. He was previously the Chief Consulting Application Scientist (Africa) for the Advanced Chemistry Development Inc., Toronto, Canada.

Contents

Preface

In prelude, the purpose of this book is to point out the increasing role of chalcogen elements in multidisciplinary fields such as biogeochemistry, material science, and biochemistry. For these reasons, the book titled 'Chalcogen Chemistry' is structured on a multidisciplinary approach by compiling chapters from various contributing authors. The chapters focus on topics ranging from reactivity to magnetic properties of chalcogens, organic to inorganic (chemistry of chalcogenides), analytical chemistry (qualitative and quantitative analysis), and synthetic to structural and theoretical chemistry (development of new materials). The book is divided into six chapters, each logically and independently presented, and then concluded with links to new research perspectives as the interest in chalcogen chemistry grows. This book is a valuable source of information not only for chemists but also for materials scientists, physicists, biochemists, and other researchers including biogeochemists, with general or particular interests in chalcogen compounds. I hope the book with a multidisciplinary approach, comprised of chapters reserved for analytical, materials, biological, and supramolecular chemistry, may induce some curiosity in readers and attract them towards this branch of chemistry.

Chalcogens ('ore builders') are elements that belong to group VI-A (or group 16) in the Periodic Table. This group consists of oxygen (O), sulfur (S), selenium (Se), tellurium (Te), and polonium (Po). Geometrically, they are known as chalcophiles, which are soft elements (keeping in mind the terminology of the hard/soft acids and bases concept). Thus, chalcophiles preferably form minerals with sulfur, selenium, and tellurium. This is the reason why, for a long time, chalcogen chemistry was mostly centred on sulfur, with selenium being treated marginally compared to sulfur, whereas the study of the chemistry of tellurium was almost non-existent. However, this consideration has changed over time. Currently, any electronic search on 'SCIfinder' will return a considerable number of scholarly published papers on chalcogen chemistry. In this regard, this book is yet another contribution to ongoing efforts in the study of chalcogen chemistry.

The interest shown in chalcogen chemistry in the past few years has motivated research focused on expanding knowledge on this special class of compounds, and today driving concerted efforts towards new materials development. Chalcogen elements are everywhere in our lives. Firstly, based on their fascinating structural and bonding principles, they span entire biotic communities connecting chemistry to biogeochemistry, biochemistry, biology, food, agriculture, and also medicine, as well as pharmacology. Secondly, the intensive worldwide exploration of chalcogen elements contained in natural compounds as minerals has now shifted the interest towards the development of these new materials or compounds (in combination with metals and ligands) with higher physicochemical properties and sophisticated characteristics. Their increased demand with far-reaching economic benefits is owed to their great contribution in modern technology in applications used by a number of industries.

The novelty of this book, compared to general textbooks on chalcogen chemistry, is that principles from other fields of science, such as Analytical Chemistry, are included as a full chapter dedicated to qualitative and quantitative characterisation of chalcogenides and chalcogen materials. This chapter presents modern methods for chalcogen compound analysis and structural determination. These methods and analytical techniques include: UV-visible and infrared spectroscopy (UV/VIS and IR), thermo analysis (TA), electrochemistry (Electrochem), magnetic analysis (MA), chromatography, X-ray methods (mostly XRF, XRD, and EDX), high resolution microscopy (SEM and TEM), multinuclear NMR, computational analysis, and bioassay. The presentation of the chapters in this book is such that there is no arbitrary separation of theoretical concepts (reactivity), chemistry (organic and inorganic), and applications of chalcogens and chalcogen materials (or chalcogenides), and an attempt has been made to stress all three equally. As such, individual chapters have attempted to present the chemistry of the chalcogen(s) or chalcogenide(s), discuss their physicochemical properties (reactivities), as well as their characterisation and applications. Within the chapters, the theoretical presentation is critically discussed concurrently with emphasis upon the industrial applications in some emerging areas such as in the biomedical, semiconductors, and biogeochemical fields. Insofar as feasible, this is also done with practical examples.

The writing of a text is a humbling experience. In retrospect, the various contributing authors would agree with me that ideas in the chapters of this book could have only been sourced from a very wide source of literature, as well as from individual research outputs. So, therefore, from the very basic and ethical principle of intellectual honesty, the authors have presented what seemed to them to be the best of their ideas, as well as those of many other authors.

While all the authors have been duly cited and referenced, it is impossible to express adequately my appreciation for the contributions of all colleagues, graduate and undergraduate students, as well as secretaries, who gave so generously of their time, effort, and opinions. In particular, the help of Ms Lada Bozic in promptly assisting both me and the authors, and handling of the editing platform activities is acknowledged. I know that my attempt to achieve depth and quality in the chapter review led to a few of them being declined that may have been considered important by some readers or the authors themselves, I hope that these omissions or inadequacies are few enough so that the usefulness of the presented chapters will not be impaired.

Peter Papoh Ndibewu
Department Chemistry, Arcadia Campus
Professor, Departments of Chemistry, Chemical & Metallurgical Engineering
Tshwane University of Technology
Pretoria, South Africa

Chapter 1

Introductory Chapter: Chalcogen Chemistry - The Footprint into New Materials Development

Ndibewu Peter Papoh

Additional information is available at the end of the chapter

http://dx.doi.org/10.5772/intechopen.82542

1. Chalcogen chemistry – the footprint into new materials development

The study of chalcogen chemistry is just fascinating for two reasons. Firstly, they span the entire biotic communities connecting chemistry to many other scientific disciplines. These include biogeochemistry, biochemistry, biology, food, agriculture, and also medicine, as well as pharmacology. Secondly, the chalcogen elements known as chalcogens demonstrate extremely interesting properties forming new compounds endowed with sophisticated characteristics that are increasingly making a remarkable footprint in a new era of materials development. These two reasons have intensified worldwide exploration of chalcogen elements contained in natural compounds as minerals. Furthermore, the aforementioned reasons have motivated research focused on expanding knowledge on this special class of compounds. In the past few decades, the shift of interest has been toward the development of new materials (in combination with metals and ligands). The contribution of this field of chemistry to the development of new materials, and their impacts on the everyday life of mankind, has triggered a recent renaissance of the interconnectivity between new chemical concepts and reactivities, resulting in a multitude of multidisciplinary focused research niche areas. The unique structures and reactivity of the class of chalcogen compounds and materials [1, 2] as well as their fascinating optical [3, 4] and electronic properties [5, 6] confer onto them very wide potential applications [7–12].

Group 16 or group VI-(A) elements in the modern periodic table comprising of oxygen (O), sulfur (S), selenium (Se), tellurium (Te), and polonium (Po) are usually called chalcogens ("ore builders") [13, 14]. The most valuable fundamental characteristic of these elements is their versatile nature to form new bonds and new compounds [15–17]. In this regard and with respect to the periodic law, the properties of the chemical elements in the chalcogen group are not arbitrary, but depend upon the structure of the atom and vary with the atomic number in a systematic way [13, 14]. In a geochemical sense, the chalcogens are referred to as soft elements, thus, also named chalcophiles [18, 19].

Geometrically, chalcophiles are the soft elements (one must keep in mind the terminology of the hard/soft acids and bases concept) [13, 20]. The modern term "chalcogen" is a derivation of two Greek words chalcos or "khalkÓs," meaning "ore formers" [19]. This word also means copper (and had also been used for bronze/brass or coin during the ancient time). One valid reason is that they are found in copper ores. The second word is a latinized Greek word "genēs" (meaning born or produced). These two words gave birth to the coined term "khalkos-genes" or to today's modern term chalcogens [21, 22]. Earlier on, chalcophiles were thought to be electron rich [13]. They are described as the low-valent elements of the Group 16 elements (O, S, Se, Te, and Po) or particularly electronegative elements and often bond to the functional groups of organic molecules [13] and/or are found as electron donors for the transition metals [23, 24]. This property has created yet another research area of interest, referred to as electrophilic and polycationic chalcogen chemistry [25]. Generally, the electron configuration of the chalcogens is represented as ns^2np^4, with two electrons short of a filled valence shell [26, 27]. The properties of O, S, Se, Te, and Po are presented in **Table 1**.

On the one hand, while chalcophiles preferably form minerals with sulfur (**Figure 1**), selenium, and tellurium, on the other hand, their unique properties in forming new compounds and materials have created a paradigm shift in research interests focusing on the deep understanding of the chalcogens and their resultant compounds also known as chalcogenides.

A great variety of technological applications are now found for the many compounds of the elements of chalcogens [28–34]. This textbook emphasizes on this shift from not only fundamental chalcogen chemistry but also on the new shift to new chalcogen materials synthesis, characterization, and testing. These trends also include fundamental research works in other fields like those in coordination chemistry focusing on a large variety of ligands [20, 23]. This subfield explores ligands containing chalcogen atoms potentially suitable for the preparation of new precursors for metal-chalcogenide vapor deposition [7]. Two examples of these macromolecules with biological activity are metallo-enzymes containing multi-metallic centers as active sites for a number of catalytic reactions and chalcogen-containing mixed macrocycles with signaling functional groups as sensors for heavy metal-ion recognition [16].

The literature [35–37] provides a variety of chalcogen compounds and materials or chalcogenides, though not exhaustive. In naming a few, it is worthwhile mentioning the class of charge-transfer salts of chalcogen-rich molecules such as tetrathiafulvalene (TTF), 1,2-dichalcogenolenes, or the variegated area of metal chalcogenides, which are all of topical interest [38]. The details of many other interesting compounds are provided in the other chapters of this book. In the field of materials and their potential applications and as an illustration, published research results and discussion span from n-dimensional molecular conductors to magnets, to bistable switchable materials, and to non-linear optical (NLO) materials [39].

This textbook discusses chalcogen chemistry from a multidisciplinary approach which ranges from synthetic to structural and theoretical chemistry, organic to inorganic and

Property	Oxygen	Sulfur	Selenium	Tellurium	Polonium
Atomic mass (amu)	16.00	32.07	78.96	127.60	209
Atomic number	8	16	34	52	84
Atomic radius (pm)	48	88	103	123	135
Atomic symbol	O	S	Se	Te	Po
Density (g/cm^3) at 25°C	1.31 (g/L)	2.07	4.81	6.24	9.20
Electron affinity (kJ/mol)	−141	−200	−195	−190	−180
Electronegativity	3.4	2.6	2.6	2.1	2.0
First ionization energy (kJ/mol)	1314	1000	941	869	812
Ionic radius (pm)†	140 (−2)	184 (−2), 29 (+6)	198 (−2), 42 (+6)	221 (−2), 56 (+6)	230 (−2), 97 (+4)
Melting point/boiling point (°C)	−219/−183	115/445	221/685	450/988	254/962
Normal oxidation state(s)	−2	+6, +4, −2	+6, +4, −2	+6, +4, −2	+2 (+4)
Product of reaction with H_2	H_2O	H_2S	H_2Se	None	None
Product of reaction with N_2	NO, NO_2	None	None	None	None
Product of reaction with O_2	—	SO_2	SeO_2	TeO_2	PoO_2
Product of reaction with X_2	O_2F_2	SF_6, S_2Cl_2, S_2Br_2	SeF_6, SeX_4	TeF_6, TeX_4	PoF_4, $PoCl_2$, $PoBr_2$
Standard reduction potential (E^0, V) ($E^0 \rightarrow H_2E$ in acidic solution)	+1.23	+0.14	−0.40	−0.79	−1.00
Type of oxide	—	Acidic	Acidic	Amphoteric	Basic
Valence electron configuration*	$2s^22p^4$	$3s^23p^4$	$4s^24p^4$	$5s^25p^4$	$6s^26p^4$

*Configuration shown in this table does not include filled d- and f-subshells.
†Cited values for the hexa-cations are for six-coordinate ions and these values are usually estimated.

Table 1. Selected properties of the Group 16 elements [26].

analytical chemistry [40], biological to material science and supramolecular chemistry. As such, it is suitable for scientists, technologists, students, as well as for those whose major interest is chalcogen chemistry, with particular interests in the chalcogen compounds and materials [41, 42]. This introductory chapter is followed by four chapters. The presentation of the chapters in this book is such that there is no arbitrary separation of theoretical concepts (reactivity), chemistry (organic and inorganic), and applications of chalcogens and chalcogen materials (or chalcogenides) while an effort has been made to stress all three grouping of ideas and facts equally. As such, in individual chapters, the authors have attempted to present the chemistry of the chalcogen(s) or chalcogenide(s) with enough discussion focusing on their structural bonding and reactivities in relation to their characterization and applications. In all the chapters, practical examples are provided insofar as possible in order to illustrate ideas.

Figure 1. A crystalline sulfur deposit. This sulfur deposit is located around a volcanic vent in Kilauea Crater, Hawaii, USA [14]; sulfur is very different from oxygen, and not very abundant; elemental sulfur is usually found in rock formations overlying salt domes, often accompanying petroleum deposits [26].

Author details

Ndibewu Peter Papoh

Address all correspondence to: ndibewup@tut.ac.za

Department of Chemistry, Faculty of Science, Tshwane University of Technology, Pretoria, South Africa

References

[1] Garcia-Vivo D, Ramos A, Ruiz MA. Cyclopentadienyl and related complexes of the group 6 elements having metal–metal triple bonds: Synthesis, structure, bonding and reactivity. Coordination Chemistry Reviews. 2013;**257**:2143-2191

[2] Boere RT, Roemmele TL. Comprehensive Inorganic Chemistry II: Chalcogen–Nitrogen Radicals. Lethbridge, AB, Canada: University of Lethbridge; 2013. pp. 376-406. DOI: 10.1016/B978-0-08-097774-4.00117-0

[3] Alekberov RI, Isayev AI, Mekhtiyeva SI, Fábián M. Local structures and optical properties of As-Se-Te(S) chalcogenide glasses. Physica B: Condensed Matter. 2018;**550**:367-375

[4] Fouad SS, Ammar AH, Abo-Ghazala M. The relationship between optical gap and chemical composition in Sb_xSe_{1-x} system. Physica B. 1997;**229**:249-255

[5] Belghit R, Belkhir H, Kadri MT, Heciri D, Ahuja R. Structural, elastic, electronic and optical properties of novel antiferroelectric KNaX (X = S, Se, and Te) compounds: First principles study. Physica B: Condensed Matter. 2018;**545**:18-29

[6] Salam MMA. First principles study of structural, elastic and electronic structural properties of strontium chalcogenides. Chinese Journal of Physics. 2018, In press, accepted manuscript. Available online 6 November 2018

[7] Manivannan R, Victoria SN. Preparation of chalcogenide thin films using electrodeposition method for solar cell applications—A review. Solar Energy. 2018;**173**:1144-1157

[8] Dai F, Shen J, Dailly A, Balogh MP, Cai M. Hierarchical electrode architectures for high energy lithium-chalcogen rechargeable batteries. Nano Energy. 2018;**51**:668-679

[9] Kim WH, Nguyen VQ, Shaw LB, Busse LE, Sanghera JS. Recent progress in chalcogenide fiber technology at NRL. Journal of Non-Crystalline Solids. 2016;**431**:8-15

[10] Shubha GN, Tejaswini ML, Lakshmi KP. Advanced material for newer applications. Materials Today: Proceedings. 2018;**5**(1) Part 3

[11] Lin C, Rüssel C, Dai S. Chalcogenide glass-ceramics: Functional design and crystallization mechanism. Progress in Materials Science. 2018;**93**:1-44

[12] Shiryaev VS, Churbanov MF. Recent advances in preparation of high-purity chalcogenide glasses for mid-IR photonics. Journal of Non-Crystalline Solids. 2017;**475**:1-9

[13] Dube JW, Ragogna PJ . Comprehensive Inorganic Chemistry II. Low-Coordinate Main Group Compounds—Group 16. London, ON, Canada: Western University; 2013. pp. 624-647. Available from: 10.1016/B978-0-08-097774-4.00124-8 [Accessed: November 12, 2018]

[14] LibreTexts™ Chemistry. 2013. Available from: https://chem.libretexts.org/LibreTexts/University_of_Missouri/MU%3A__1330H_(Keller)/22%3A_Chemistry_of_the_Nonmetals/22.06%3A_The_Other_Group_6A_Elements%3A_S%2C_Se%2C_Te%2C_and_Po [Accessed: November 13, 2018]

[15] Ghorai P, Brandão P, Bauzá A, Frontera A, Saha A. Anion-reliant structural versatility of novel cadmium(II) complexes: Synthesis, crystal structures, photoluminescence properties and exploration of unusual O···S chalcogen bonding involving thiocyanate coligand. Inorganica Chimica Acta. 2018;**469**:189-196

[16] Mahmudov KT, Kopylovich MN, Guedes da Silva MFC, Pombeiro AJL. Non-covalent interactions in the synthesis of coordination compounds: Recent advances. Coordination Chemistry Reviews. 2017;**345**:54-72

[17] Bessières B. Functions containing a chalcogen and any group other than a halogen or a chalcogen reference module in chemistry. Molecular Sciences and Chemical Engineering. 2014

[18] Werner F. A second note on the term 'Chalcogen'. Journal of Chemical Education. 2001. Available from: http://jchemed.chem.wisc.edu/Journal/Issues/2001/Oct/abs1333_1.html. [Accessed: November 11, 2018]

[19] Greenwood NN, Earnshaw A. Chemistry of the Elements. New York: Pergamon Press; 1984

[20] Ephritikhine M. Molecular actinide compounds with soft chalcogen ligands. Coordination Chemistry Reviews. 2016;**319**:35-62

[21] Cotton FA, Geoffrey W. Advanced Inorganic Chemistry: A Comprehensive Text. 2nd ed. New York: Interscience Publishers; 1966

[22] Lide DR, Frederikse HPR, 2005. editors. CRC Handbook of Chemistry and Physics: Special Student Edition. 77th ed. Boca Raton, FL: CRC Press; http://webdelprofesor.ula.ve/ciencias/isolda/libros/handbook.pdf [Accessed: November 11, 2018]

[23] Dipak Kumar Dutta DK, Biswajit D. Potential rhodium and ruthenium carbonyl complexes of phosphine-chalcogen (P-O/S/Se) donor ligands and catalytic applications. Coordination Chemistry Reviews. 2011;**255**(15-16):1686-1712

[24] Sberegaeva AV, Watts D, Vedernikov AN. Oxidative functionalization of late transition metal-carbon bonds. Chapter in Advances in Organometallic Chemistry. May 2017. https://www.researchgate.net/publication/316972402_Oxidative_Functionalization_of_Late_Transition_Metal-Carbon_Bond [Accessed: November 11, 2018]

[25] Saberinasab M, Salehzadeh S, Solimannejad M. The effect of a strong cation···π interaction on a weak selenium···π interaction: A theoretical study. Computational and Theoretical Chemistry. 2016;**1092**:41-46

[26] National Science Foundation (NSF). The California State University affordable learning solutions, and merlot. 2018. LibreTexts™ Chemistry. info@libretexts.org. Available from: https://chem.libretexts.org/Core/Inorganic_Chemistry/Descriptive_Chemistry/Elements_Organized_by_Block/2_p-Block_Elements/Group_16%3A_The_Oxygen_Family_(The_Chalcogens). [Accessed on January 12, 2018]

[27] Devillanova FA, du Mont W-W, editors. Handbook of Chalcogen Chemistry (New Perspectives in Sulfur, Selenium and Tellurium). 2nd edition, Vol. 1. The Royal Society of Chemistry (RSC) Publishing; 2013. Available from: http://scholar.google.com/scholar?hl=en&btnG=Search&q=intitle:Handbook+of+Chalcogen+Chemistry#1%5Cnhttp://dx.doi.org/10.1039/9781849737463 [Acessed on October 1, 2018]

[28] Shpotyuk M, Kovalskiy A, Golovchak R, Shpotyuk O. Phenomenology of γ-irradiation-induced changes in optical properties of chalcogenide semiconductor glasses: A case study of binary arsenic sulphides. Journal of Non-Crystalline Solids. 2018;**498**:315-322

[29] Shieh M. Ternary copper-incorporated group 8 (Ru or Fe) carbonyl chalcogenide complexes and polymers: From syntheses to applications. Journal of Organometallic Chemistry. 2017;**849-850**:219-227

[30] Fraenkl M, Frumarova B, Podzemna V, Slang S, Wagner T. How silver influences the structure and physical properties of chalcogenide glass (GeS2)50(Sb2S3)50. Journal of Non-Crystalline Solids. 2018;**499**:412-419

[31] Xiao JR, Yang SH, Feng F, Xue HG, Guo SP. A review of the structural chemistry and physical properties of metal chalcogenide halides. Coordination Chemistry Reviews. 2017;**347**:23-47

[32] Gago AS, Luo Y, Alonso-Vante N. Chalcogenide electrocatalysts for energy conversion fuel cell. Encyclopedia of Interfacial Chemistry. 2018:419-445

[33] Lucas J, Troles J, Zhang XH, Boussard-Pledel C, Bureau B. Glasses to see beyond visible. Comptes Rendus Chimie. 2018;**21**(10):916-922

[34] Ji W, Chu X, Zhang S, Wang D, Ma Y. Active basal plane in ZT-phased MX2 (M = Mo, W; X = S, Se, Te) catalysts for the hydrogen evolution reaction: A theoretical study. International Journal of Hydrogen Energy. 2018;**43**(42):19432-19437

[35] Chand S, Thakur N, Katyal SC, Barman BP, Sharma P. Recent developments on the synthesis, structural and optical properties of chalcogenide quantum dots. Solar Energy Materials and Solar Cells. 2017;**168**:183-200

[36] Kuchta MC, Parkin G. Terminal chalcogenido complexes of Group 13 and 14 elements. Coordination Chemistry Reviews. 1998;**176**:323-372

[37] Taylor R. Progress in the understanding of traditional and nontraditional molecular interactions. Comprehensive Medicinal Chemistry III. 2017:67-100

[38] Papavassiliou GC, Terziz A, Delhaes P. Tetrathiafulvalene (TTF), 1,2-dichalcogenolenes and their conducting salts. In: Nalwa HS, editor. Handbook of Organic Molecules and Polymers. Vol. 1. Cahrge-Transfer Salts. Fullerences and Photoconductors. Chichester, New York, USA: John Wiley & Sons Ltd; 1997

[39] Zhang J, Sun Z, Zhang W, Ji C, Liu S, Han S, et al. Hydrogen-bonded switchable dielectric material showing the bistability of second-order nonlinear optical properties. Crystal Growth & Design. 2017;**17**:3250-3256

[40] Kelly PF, King RSP. Catenated sulfur compounds. In: Jan R, Kenneth P, editors. Comprehensive Inorganic Chemistry II. 2nd ed. Vol. 1. 2013. The Netherlands: Elsevier B.V.; 2018. pp. 7544. ISBN: 978-0-08-096529-1

[41] Véronique Sousa V. Chalcogenide materials and their application to non-volatile memories. Microelectronic Engineering. 2011;**88**(5):807-813

[42] Tiwari AP, Kim D, Kim Y, Prakash O, Lee H. Highly active and stable layered ternary transition metal chalcogenide for hydrogen evolution reaction. Nano Energy. 2016;**28**: 366-372

Modern Analytical Chemistry Methods for Chalcogen Materials Analysis and Characterization

Surjani Wonorahardjo, Fariati Fariati and
I Wayan Dasna

Additional information is available at the end of the chapter

http://dx.doi.org/10.5772/intechopen.77989

Abstract

Analytical methods are needed to elucidate modern and complex compounds as well as to describe their physical properties. The underlying principles of chalcogen chemistry as well as the natural abundance of chalcogen elements are the base of building many biological substances, including sophisticated materials for future applications. Thus, the need for modern and state-of-the art analytical methods and techniques to characterize them, is obvious. In this chapter, challenges in analytical methods for chalcogen compounds and materials, as well as some examples of natural or synthesized materials or their combinations, including biomaterials, are discussed. Modern methods for chalcogen compounds analysis and structural determination discussed include: UV-Visible and infrared spectroscopy (UV-Vis and IR), thermo analysis, electrochemistry, magnetic analysis, chromatography, X-ray methods (mostly XRF, XRD, and EDX), high-resolution microscopy (SEM and TEM), multinuclear NMR, computational analysis, and bioassay. Also, the historical background and nature of chalcogen elements, including reactivity and magnetic properties as well as thermal behavior, common compounds of chalcogen elements: organic and inorganic materials, complex chalcogen materials, will be briefly discussed.

Keywords: analytical methods, chalcogen materials, characterization, application

1. Introduction

People living in this scientific era might not realize that chalcogen elements are everywhere in our lives. Especially oxygen which is essential for living, and then sulfur and selenium which are so close to the human life, as they are incorporated or form parts of essential and useful compounds and materials. Tellurium and polonium might not be inbound to life, but they are utilized for some purposes according to the new properties and attributes of material, as nanoselenium with their high active surface area and catalytic efficiency [1, 2]. The interest for chalcogen materials is growing rapidly, as seen through SCIFinder [3]. Combinations with other elements make the compounds important for metabolism or for various dedications. The investigation on possibilities of chalcogen elements in reaction has been thoroughly discussed from the internal chemical properties of the elements [4]. Availability of pair and lone-pair electrons offers so many possibilities of combination, naturally and synthetically. There has also been a description of the stability of organo-chalcogen compounds by intermolecular coordination because of their role in chiral induction [5].

Secondary metabolites and study on metabolomics have dominated the area of biology, food, agriculture, and also medicine, as well as pharmacy. One good example, is the chemistry of garlic, in which the sulfur-containing compounds called Allicin dominated the most beneficial properties of garlic as an antioxidant, antimicrobial and antifungal actions [6]. More similar analysis on *Allium* family revealed that mostly sulfur compounds were responsible for the antioxidant and antimicrobial properties [7]. The phytochemical profiles describe the benefits of sulfur-containing compounds to the good bioproperties which have good prospect in the future. Besides sulfur, selenium plays a preponderant role in cellular metabolism and becomes an essential element in enzyme in protecting the body against oxidative damage as well as many other functions [1]. Selenium is needed in life in certain amount; otherwise, it can be toxic, as sometimes it is called by double-edged sword element [8]. This element is also being well investigated by scientists in relation with its bioavailability in the environment [1, 2]. Originally, selenium is present in rocks, water, and soil, and it has some common isotopes too [9], though it naturally occurs in biological cycles in the environment, including the biochemical and food cycles. Selenium is one of the metalloid essential minerals in living things, whereby its deficiency, in soil and crops can cause certain metabolic disease such as the Kashin-Beck disease, commonly prevalent in the Tibetan plateau [10]. In short, the exploration of chalcogen elements contained in natural compounds as minerals is still intensive worldwide, while a shift toward agricultural use is currently underway [1, 9, 11].

The new materials in combination with metals and ligands with higher specification and sophisticated preparation are in demand due to great contribution for modern technology. Selenium, for example, can be immobilized through reaction to form selenium metal-humic ternary complexes or incorporated in carbohydrates as well [9]. More related to life, thio-compounds cannot be uncounted for since they are all over the synthetically chemicals through the ligands or thiolates as well as selenolates [12]. Synthetic compounds with metals like Fe by chalcogen arsenide [13] or chalcogen containing iron-carbonyl clusters [14] are already done. Chalcogen elements as dopant in silicon layers [15] or as organo-selenium or tellurium cations [16] are also analyzed by spectroscopy methods. Tellurium is a bigger chalcogen element and tends to have isotopic abundance, from which some are stable and some unstable [17].

There are several reports on chalcogen material synthesis with various synthetic and analytical methods, available in the Handbook of Chalcogen Chemistry [3]; however, with the emergence of new and complex compounds, analytical methods development is undoubtedly becoming an endless effort in the field of chalcogen chemistry. How the new materials are explored and investigated and also how the applications are being established and confirmed are both the new challenges of analytical chemistry. In the future, the chalcogen materials cannot be discussed within the scope of chemistry only, but must be broaden in the context of science in a more general knowledge.

2. Nature of chalcogen compounds and chalcogen materials

Chalcogens are elements which belong to the group VI-A (or group 16) in the periodic table, which consists of oxygen (O), sulfur (S), selenium (Se), tellurium (Te), and polonium (Po). Chalcogens are the basic elements of chalcogenides compounds, where chalcogens are combined with electropositive elements, organic radicals, in natural secondary metabolites and even in macromolecules such as enzymes and proteins [1, 18, 19]. All compounds with oxygen can be essentials and contribute most of the chalcogen materials. The materials can be as simple as silica, as an example of oxygen compounds was made for some important application [20] or macromolecule like cellulose or its derivates [21] which is applicable in daily life and will be so still in the future, to a more complicated complexed compounds [22–26] or advanced materials [27–29] which are made for the sake of science and method development. Abundant materials are naturally in existence with oxygen, and also a lot is being derived from synthesis in the laboratory [20–26, 30].

Some literature exclude oxygen from the chalcogenides discussion as oxygen appears in almost all materials [3]. Oxygen and sulfur are contained in so many secondary metabolites in living things and also in bigger molecules in tissue [6, 18, 31–34]. The process of changing compounds during food treatment enables one to establish new natural resources for modern herbal medicine for cancerous treatment [7, 35–39]. To be able to examine the biomolecular structure and medicinal properties of such compounds in detail, new analytical methods and also high-resolution instrumentation would play important roles [40–42]. Of course that pretreatment prior to analysis will be taken into account [31, 41, 43]. Extraction process which is needed before the analysis was discussed thoroughly in the prominent book of Harborne [44]. Many sophisticated new materials are made and dedicated for use in medicinal area [35, 38, 45] or smart electronic materials [46], materials from nonliving beings [3, 47, 48], nanomaterials and thin films and materials for certain applications which needs special characterizations [49, 50]. The fast development of chalcogen materials cannot be separated from the development of analytical methodologies.

3. Chalcogen bonds

Most topics on chalcogen compounds and materials are related to real chemical bonding, which are either ionic or covalent bonding in one molecular building. Fewer discussions describe the weak intermolecular forces that bind molecule or macromolecule together for certain purpose. Chalcogen bonds are important in many intermolecular interactions which in turn determine the configuration and designs of bigger biomolecules [51]. This type of bonding can be grouped as dipole–dipole interaction. In the field of biochemistry, it has much impact due to its role in bigger molecular mobility. The chalcogen bonds also form eclipse (*cis*) and staggered (*trans*) configurations, enable chalcogen elements to develop many types of materials for specific application, from real crystals to real amorphous substances.

The intra-molecular chalcogen bonds were obtained from the X-ray diffraction results and quantum chemical calculations, such as in thioindirubin [52], indicating advanced methodology in modern analysis of big molecules. Chalcogen bonds were also studied with the aid of computational programs, as the development of old theories, especially in the debates around the energetic significance and physicochemical origins of the so-called class σ-hole interaction [53], continue to fuel scientific discussions. These approaches have served as important steps toward the synthesis, analysis, and designing of new materials [51].

4. Challenges in analytical methods for chalcogen compounds and materials

Natural chalcogen compounds which are present in certain matrix in perfect blends, and separation procedure must be conducted first, followed by purification prior to analysis. There are several parameters must be taken into account for the correctness of measurements. In the extraction of sulfur-containing compounds in plants, one would normally use phytochemical procedures [44] by considering the separation of constituent analytes according to the similarities of the compounds.

The development of new materials from chalcogen compounds due to the functionality of the materials must be supported by a better analytical methodology. When materials are dedicated to a specific purpose, then the proof to that claim might be assigned analytically too. In this case, there are two types of analytical chemistry for the assignment of chalcogen materials. In short, the need of analytical method can cover four types of analysis:

1. The need of pretreatment method prior to analysis, for the natural compounds in natural matrix.
2. The need of methods for characterization of the chalcogen compounds, as well as new materials, in practice and theoretically.
3. The methods to describe the application for the new materials which include several other fields of disciplines.
4. The need of methods validation involving more than one method and instrumentation.

All of the human efforts in the laboratory as well as in computational analysis are based on the four types of analytical objectives listed above. The emphasis must depend on the purpose or in one segment of a longer process and all point of views can count. Therefore, any bigger steps can be started from chemistry discussion and developed into a wider investigation perspective.

5. Methods for chalcogen materials characterization

Characterization method is the backbone of chalcogen chemistry material description, as it always accompanies the explanation of material properties [54–57]. There are still some divisions in chalcogen material characterization, which includes the analysis for the main material itself and characterization of the impurities. The presence of impurities will decrease the quality to some extent [58, 59]. Notwithstanding, there have been too few reports discussing the impurities aspect if not related to their main functionalities.

Basic spectroscopy methods, especially X-ray methods, are the main tool for material characterization, including chalcogen materials. The methods are based on incoming X-ray beam that undergoes some natural phenomena like absorption, emission, fluorescence, and diffraction, then scattering with many possibilities to explore the chemical composition and properties of the sample. The crystallinity of materials can be derived from the X-ray diffraction patterns and the crystal database from instrument companies. In this case, the X-ray penetrates through the materials, and a number of particles can be expected to be oriented in such a way as to fulfill the Bragg's law. Almost all crystalline compounds analysis rely on XRD spectra, such as analysis of metal complexes of metal-thiourea and metal phenyl-thiourea [60, 61] after several steps of synthesis, to determine the coordination sphere on the metal Zn(II), Co(II) and Cu(II) and the possible crystal structures. More study of the spectra confirmed the shape of crystalline compounds together with UV-Visible and infrared as well as magnetic susceptibility measurement. In other synthesis the X-ray spectra were used to calculate reactive tendency as well as shape of molecules [23–25] besides characterization. Another X-ray technique is the energy dispersive analysis by X-rays (EDX) which intensity is proportional to the amount of the elements. The method is commonly combined with scanning electron microscopy (SEM) to get pre-experimental data, before the variables are given [22, 24, 26] or to characterize and give elemental confirmation [62].

Nuclear magnetic resonance for solid sample can be powerful to characterize chalcogen materials. It is based on the impact of radiofrequency irradiation on specific nuclei in certain field strength of the magnet (FT-NMR), causing the nuclei to spin resulting in resonance frequency which is an indication of the atom (e.g., ^{77}Se and ^{125}Te) [28]. NMR for chemist such as ^{1}H or ^{13}C NMR is usually the most important method for chemical structure elucidation. However, material scientists need solid-state NMR with its magic angle spinning (MAS). Moreover, when other nuclei of resonance are used, one must swing the magnetic field according to selected NMR active probe nuclei, such as ^{77}Se and ^{125}Te NMR [28, 29] to describe and confirm newly synthesized octahedral coordination compounds. NMR method assisted the description of how chalcogen elements (Se and Te) that can replace halogen as inner ligands in forming cluster cores in octahedral cluster complexes. This action reduces the symmetry and makes distortion on the metallic cluster as well as their isomers and of course changes the properties of the whole materials. Proton and (^{1}H) carbon (^{13}C) NMR of complex protein molecules were employed to investigate *Se*-(2-aminoalkyl)selenocysteines as biochemical redox agents [63]. In this case, chalcogen-biochemical substance was investigated through the behavior of its protons and carbons. The similar proton, carbon as well as ^{77}Se spectra recorded, was also the main method of metal chalcogenides characterization synthesized from single source molecular precursor, besides X-ray diffraction [64]. This NMR method extended to confirm the presence of chalcogen atoms in the complex molecular building through ^{195}Pt NMR. There was also a good NMR result which suggested the coordination of thiones to Zink(II) although the sulfur (S) atoms indicated an up-field shifting of = C=S resonance of ^{13}C NMR as well as a downfield N-H resonance in ^{1}H NMR [65]. Similar analysis was done for the characterization of complex coordination compound using thiourea derivative for biochemistry research purposes [66]. The proton NMR was also used to analyze the stability of intermolecular coordination as one good aspect is organo-chalcogen compound [5]. One can see that NMR is very useful in describing molecular properties and mobility.

Thermogravimetric analysis is a method in which changes in physical and chemical properties of materials are measured as a function of increasing temperature (with constant heating rate) or as a function of time (with constant temperature and mass loss). The important method of differential thermal analysis (DTA) one in which the mass changes is related to specific heat capacity. The increase of temperature is programmed to be linear and the heat flow to both sample and reference. Solid decomposition will occur due to phase changes, and this process can be endothermic or exothermic. One example of the methods on chalcogen compound was reported in the discussion of TGA of zinc and cadmium thiolate and selenolate complexes that showed the formation of metal sulfide (MS) and metal selenides (MSe) [M = Zn, Cd], while the mercury complexes showed complete weight loss in this temperature range [5]. TGA is mentioned as a good method to characterize metal chalcogenides [67] and has become a key analytical information generating technique together with X-ray diffraction data to validate the formation on chalcogen arsenide clusters in the iron with carbonyl functional groups [13]. In studying the degradation of palladium thiolate and selenolate, TGA was also used, to confirm the formation of Pd_4S and $Pd_{15}Se_{17}$ which was then characterized by XRD and EDX [64].

Optical microscopy is used after the synthesis steps of chalcogen materials, which aids the visual characterization of materials. This method is improved continuously and became the earlier stage of today's electron microscopy. Reflection mode of the instruments is preferable, and this method is called episcopic light differential interference contrast (DIC) microscopy, which enables imaging of polymer, glasses, semiconductor, metals and minerals sample with various reflective properties. DIC microscopy also has its limitation, as it gives experimental uncertainties during measurement, which is discussed in several numerical methods to minimize them [57].

Scanning electron microscopy (SEM), which is a technique used for a better description of materials' surface textures up to nanometer scale, is a good way of visualizing chemistry by secondary electrons. The ability to focus the extremely small incident wavelength of the energetic electrons to resolve object in extraordinary spatial resolution, makes the method popular for nanotechnology's purposes. Electrons are scattered very intensively compared to X-rays in both elastic and inelastic ways for both organic and inorganic materials, in dimensions less than 1 nm. While transmission electron microscopy (TEM) is a similar microscopy electron, but the image is formed from the passage of some electrons passing through thin sliced samples together with the elastic and inelastic scattering of the electrons. Thicker samples result in decreasing energy of the electron beams and increasing the scattering as well as complexity from the bigger distribution of energy and at the end declined resolution is obtained. Most of the synthetic chalcogen materials or metal chalcogenides are firstly being visualized with the aid of SEM and TEM before any other theoretical modeling or use of methods for analysis [15, 20, 24, 50, 56, 62, 68–70].

Dispersive infrared (IR) spectroscopy has been in use and became more popular with the modern nondispersive Fourier transform-infrared (FT-IR) systems to probe the presence of certain functional groups. From the energy point of view, vibrational frequencies are the base of most analyses, and rotational frequencies also count. Raman spectroscopy, on the other hand, from very different principles of spectroscopy, the scattered intensity of the absorbed energy informs the same energy absorbed by vibration. Sulfur is observable by infrared spectroscopy [66, 71, 72] by their vibrational modes, especially stretching and bending vibrational modes in solid, liquid, or gaseous phases. Fingerprint region is also important. Bulk characterization by IR was employed to analyze synthetic compounds to prove the presence of thionyl vibrational mode (ν(C-S)) with frequency band shifts to lower values after coordination with metallic atoms [66]. Similar recording of infrared spectra were found in metal complexes of thiourea and phenylthiourea crystals [60, 61]. The C-H stretching of the components overlapped with N-H stretching of the thiourea, but both can be differentiated since N-H is not directly involved in bond formation with the metals. The metal ion complexation on the ligands is more pronounced on N-C and C=S bond which is shifted after the complex is formed. It is also confirmed that the phenylthiourea is coordinated to metal *via* the sulfur with a reduction in π-electron density of the C=S bond.

Surface characterization modes use additional probes such as attenuated total reflectance (ATR) or diffuse reflectance infrared Fourier Transformed spectroscopy (DRIFT) [9] or reflection absorption infrared spectroscopy (RAIRS). Sorption study of selenium(IV) solution on natural zeolites was done by infrared spectroscopy [73]. In this case, pH and concentration of sodium selenite solution onto shabazite, analcime, stilbite, mesolite from volcanic fields were studied. Some new absorption bands from Se-O as well as Se-O-Se bridges were observed, different from original infrared spectra recorded before. The strongest changes due to the highest pH of sodium selenite were the shifted absorption of tetrahedral Al-Si-O of the natural zeolite framework downfield in alkaline situation and another band appeared that confirms the absorption state of the ions. The partial desilylation of zeolite in alkaline medium as well as dealumination of zeolites occurring in acidic solution were observed in the infrared spectrum. In thin film form materials, near infrared (NIR) analysis was used to compare the photo-response of silicon doped with Se and Te via laser irradiation [15]. In this case, surface morphology and optical properties were accessed by NIR spectroscopy, as well as the stability of Si-chalcogen interaction. In other discussion, the vibration-rotation spectrum informs the bond length of the molecules being investigated [71] and process chemistry can be followed. A very minute detail of absorption energy can be useful, making this method valuable from time to time during synthesis, for calculating and determining the crystal building of the structure. Moreover, some workers have used IR spectroscopy to complement the computational calculation of new inorganic complex cluster with chalcogen elements [29].

Most of organic and inorganic compound or ions adsorb radiation in the ultraviolet and visible region (UV-Vis) (180–750 nm). Part of chalcogen materials also produces electronic spectra that show shape of molecules or crystal as a result of the frequency absorption bands from ligands, especially for the bands near the visible region as expected [60, 61]. Furthermore, the electronic transition in *d*-orbitals also provides strong evidence for complex compounds containing transition metals. Examples of similar complex metal chalcogenides follow the same principles for different shapes of molecules, together with analysis of magnetic susceptibility, which suggests the shape of environment of the central metal ions with the presence of chalcogen ligands.

One other important analytical method for both characterization as well as application of chalcogen materials is electrochemistry [74]. The role of electrochemistry in synthesis, development, as well as characterization, up to applications, is obvious. This method is based on electron transfer in chemical reactions, in which metals have the most possible elements for electron storage systems. In photo-electrochemical systems, in which electron from the reaction is to be stored as energy or used for the next reaction.

Characterization of magnetic properties is also important in the study of chalcogen materials. Before, vibrating sample magnetometer was used to get information about the magnetization of samples when vibrated in a uniform magnetizing field. Magnetization is therefore induced, the product of magnetic susceptibility and the applied magnetic field provide chemical information of the materials. The specific techniques include: magnetic separation, magnetic spectroscopy, magnetic susceptibility measurement, magneto-relaxometry, magnetic particle spectroscopy, and rotating magnetic field. Some magnetic properties can be changed due to chalcogen substitution to metal iron complexed compounds [55, 75].

Dynamics in chalcogen materials is also trending in the field since it is crucial to describe the desired properties of the materials. In addition, the dynamics of materials are now core in understanding conductivity and diffusivity of the materials [76]. Materials with ion dynamics of different substructures enable phonon scattering process in their solid state. Actually, NMR relaxation and diffusion experiments are powerful tools used to describe molecular mobility, no matter what the nuclei probe is [77, 78]. The same method can be applied to characterize chalcogen materials as well as metal chalcogenides.

Since computer is involved in most of the modern chemical analysis, the chalcogen bonds in protein are one example in this field [79]. The computational analysis needs some unfamiliar tools and methods; however, it provides a lot of information about how molecules bind together naturally. Sulfur, selenium and tellurium are the probes of energetically favorable trends in the synthesis of chalcogen complex structures [29], following modeling by computational analysis. In addition, while there are more types of interaction occurring together in the biomolecules, which one cannot resolve them one by one, nonionic and noncovalent bonds are usually resolved by computational calculation. The intramolecular forces between sulfur and oxygen was also reported as the chalcogen bonds which is responsible for many bond formations in bigger molecules [52] is often being modeled by means of computation.

6. Methods for testing the applications of chalcogen materials

The method of applications will depend on the field of applications. The difficult part of it is to find a probe or indicator for the desired properties needed to be performed by the materials. This includes more analytical chemistry, with biological capacity or computer calculations. Suitable characterization is also essential to correlate the application and the properties of the materials.

Application of many types of chalcogen materials for environmental purposes employs infrared sensing for chalcogenides fibers [72] and also extended to other signal in infrared region can be utilized for environmental sensors. The manufacturing and testing of optical fiber sensors made from transparent chalcogen compounds for environmental can also be an alternative device for use in infrared spectroscopy [80].

For agricultural or medicinal application, bioassay is mostly used. New materials for antimicrobial properties are tested using qualitative or quantitative microbial assay [60, 61], in which pathogenic microbials were used to test the biological potentials of the compounds synthesized, as seen from disc diffusion method. The cultured microorganism in petri dishes would give clear inhibition zones around a spot of medium impregnated with stock solution of the synthesized complexes during incubation under certain conditions. Potential antibacterial activity can be further traced quantitatively. Usually, several methods are used together for the specific area of applications. Moreover, many methods can be compared one to other

Methods	Characterization	Application	Explanation
Uv visible	[60, 61]		Characterization of compounds structures.
		[68]	For the analysis of colored compound when probing SO_x and NH_3 reduction.
IR	[13, 15, 24, 66, 68, 73]		For the characterization of functional groups after reactions, characterization of chalcogen dopants on silicon. Sorption study of selenium(IV) solution on natural zeolites
		[15]	Investigation on fabrication of silicon based new material by near infrared analysis.
Luminescence	[27]		Characterization of chalcogen compounds by spectral luminescence study.
XRD	[13, 20, 21, 23, 24, 51–53, 60, 61, 73]		Determination of crystallinity of materials and its combination, for characterization of intermolecular interactions, the effect of chalcogen substitution
NMR	[2, 5, 13, 27, 28, 61–64]		Characterization of hydrogen and carbon- containing groups, also Se and Te NMR for chalcogen elements in the molecules
		[81–85]	For the mobility of small molecules in porous oxide materials, testing of chalcogen material application

Methods	Characterization	Application	Explanation
Mass spectrometer	[13]		Characterization of compounds synthesized
		[86]	For the determination of volatile compounds released from the silica oxides materials, testing of chalcogen materials application
Atomic spectroscopy	[10]		Quantitative measurement of selenium content in soil using HG atomic fluorescence spectroscopy
Optical microscopy	[18, 24, 85]		Surface texture characterization, as well as structural components of chalcogen materials.
Chromatography		[20, 86]	Separation of plant pigments on surface of silica materials, as one application for chalcogen compounds
Thermoanalysis	[3, 13, 64, 65, 80]		Characterization of chalcogen compounds, thermal clusters on chalcogen materials, formation of Zinc and cadmium sulfide and selenide, the study of degradation of palladium thiolate and selenolate
Magnetic analysis	[56, 60, 61, 75]		Characterization of magnetic nanoparticle of chalcogen materials, the effect of chalcogen substitution, suggestions for shape of complex metal chalcogenides together with electronic transition results.
Electrochemistry	[4]		Characterizing molecules based on reactions of chalcogen compounds
		[49, 69, 74]	Voltammetric measurement for application chalcogen doped Mo as electrode, photoelectrochemical solar cells with chalcogen materials
SEM	[18, 24, 48, 54, 60, 66–68]		Describing porous texture of the surface
EDX	[24, 64]		Describing the composition of elements of materials, characterization of Pd_4S and $Pd_{15}Se_{17}$
Bioassay		[60, 61]	Testing bioactivity of the chalcogen materials (thiourea).
Computation methods	[52, 53]		Calculation on S···O chalcogen bonds and modeling

Table 1. Chalcogen (including oxygen) materials analysis using available analytical techniques.

since some treatments are meant to be complementary to others. Method validation between more than one approach would be important. **Table 1** presents some available analytical methods for characterization and testing the applications.

In modern information technology, new challenges come from big data handling too. Most recent development in the handling of chemistry data, qualitative and quantitative can provide more information about the materials. Chemometrics has also become an important analytical chemistry tool in many disciplines including chemistry and other applied chemistry fields (e.g., biochemistry and bioinorganic chemistry). The data analysis gives trends, and this hypothetical analysis creates continuity in the investigation. Thus, the challenges of analytical chemistry are real, yet exalting!

7. More accounts on development of chemistry now and for the future

It is clear that there is an increasing demand for modern materials for various industry applications [67]. Therefore, modern analysis is needed to ensure that the synthesis of these newly made materials complies with quality attributes and satisfy their purposes. With the emergence of new devices, information technology and materials for big data handling, demand in the field of health and the pharmaceutical sector, as well as materials for application in science and the environment, are all factors bound to accelerate research to produce more types of new materials. At this point in time, chemistry is the key in technological and engineering developments, as everything can be manipulated from molecular [87] to structural levels. Chemistry education has the perspective of chemistry contextualization to socio-scientific orientation [88]. Awareness of chemistry concepts and also biochemistry is crucial in chemistry teaching, since the chemistry content alone is not enough to shape up the scientific attitude. When technology reigns, without good attitude of the chemist behind that, then the society as well as the future as a whole is in danger. So, it is clear here, that chemistry education is important for the right technology for the benefit of mankind.

Last but not the least, material science is a new fascinating area of interest attracting more and more scientists around the world. The need of raw materials leading to earth exploration and exploitation especially in mining has shifted the natural equilibrium to some extent and in turn will move as a self-reorganization phenomenon which in the context of human and other living being is categorized as a disaster. In this case, scientist must work within the scope of ethics since any changes from structural level, from chemistry level can develop up to environmental level. The idea of green chemistry has been developed due to new awareness of imbalancing nature by human activity. Chemistry education has to be more "eco-reflexive" [89] and technology must develop the environment with responsibility.

8. Conclusion

One of the main problems analytical chemist has to face is the lack of compositional and structural information concerning chalcogens and related chemical compounds and materials. The development of efficient procedures for the synthesis, extraction, and characterization or structural determination of this class of compounds is bottleneck of each analysis, for both characterization as well as optimization and applications. Thus, the analytical protocol for such analysis usually performed with the use of a wide range of techniques, both single and hyphenated, should be designed on the basis of the need to provide required knowledge about translocation of the metal and character of its interactions with examined chalcogen materials or compounds toward establishing goal oriented method. Method development as well as validation is, therefore, crucial in shaping future technology development and application of chalcogen and chalcogenides. Analytical chemistry premised on new ideas out of human creativity is set to shape future technology for chalcogen materials.

Conflict of interest

We declare that this chapter has no "conflict of interest."

Author details

Surjani Wonorahardjo*, Fariati Fariati and I Wayan Dasna

*Address all correspondence to: surjani.wonorahardjo@um.ac.id

Chemistry Department, Faculty of Mathematics and Science, State University of Malang, Malang, Indonesia

References

[1] El-Ramady HR, Domokos-Szabolcsy É, Abdalla NA, Alshaal TA, Shalaby TA, Sztrik A, et al. Selenium and nano-selenium in agroecosystems. Environmental Chemistry Letters. 2014;**12**(4):495-510

[2] El-Ramady H, Abdalla N, Taha HS, Alshaal T, El-Henawy A, Faizy SEDA, et al. Selenium and nano-selenium in plant nutrition. Environmental Chemistry Letters. 2016; **14**(1):123-147

[3] Devillanova F. Handbook of Chalcogen Chemistry [Internet]. Vol. 2. Cambridge: RSC Publishing, The Royal Society of Chemistry; 2013. P001-588p. Available from: http://scholar.google.com/scholar?hl=en&btnG=Search&q=intitle:Handbook+of+Chalcogen+Chemistry#1%5Cn. http://dx.doi.org/10.1039/9781849737463

[4] Glass RS. Redox chemistry of sulfur, selenium and tellurium compounds. In: Woollins JD, Selenium, Laitinen RS, editors. Tellurium Chemistry [Internet]. Berlin: Springer; 2011. pp. 57-77. Available from: http://link.springer.com/10.1007/978-3-642-20699-3

[5] Mugesh G, Panda A, Singh HB. Aspects of organochalcogen (S, Se, Te) compounds stabilized by intramolecular coordination. Journal of Chemical Sciences. 2000;**112**(3):239-248

[6] Harris JC, Cottrell SL, Plummer S, Lloyd D. Antimicrobial properties of *Allium sativum* (garlic). Applied Microbiology and Biotechnology. 2001;**57**(3):282-286

[7] Khalid N, Ahmed I, Latif MSZ, Rafique T, Fawad SA. Comparison of antimicrobial activity, phytochemical profile and minerals composition of garlic *Allium sativum* and *Allium tuberosum*. Journal of Korean Society for Applied Biological Chemistry. 2014;**57**(3):311-317

[8] Fordyce FM. Selenium deficiency and toxicity in the environment. In: Selinus O. editors. Essentials of Medical Geology. Revised ed. Dordrecht: Springer; 2013

[9] Fernández-Martínez A, Charlet L. Selenium environmental cycling and bioavailability: A structural chemist point of view. Reviews in Environmental Science and Biotechnology. 2009;**8**(1):81-110

[10] Wang J, Li H, Li Y, Yu J, Yang L, Feng F, et al. Speciation, distribution, and bioavailability of soil selenium in the Tibetan plateau Kashin-beck disease area—A case study in Songpan County, Sichuan Province, China. Biological Trace Element Research. 2013; **156**(1-3):367-375

[11] Van Geem M, Harvey JA, Cortesero AM, Raaijmakers CE, Gols R. Interactions between a belowground herbivore and primary and secondary root metabolites in wild cabbage. Journal of Chemical Ecology [Internet]. 2015;**41**(8):696-707. Available from: http://link.springer.com/10.1007/s10886-015-0605-7

[12] Frolov KA, Dotsenko VV, Krivokolysko SG, Kostyrko EO. Novel [4 + 2] cycloaddition reaction of aryl-methylidenemalononitriles to unsaturated chalcogen amides. Synthesis, structure, and properties of triethylammonium 3,5,5-tricyano-1,4,5,6- tetrahydropyridine-2-selenolates and -thiolates. Chemistry of Heterocyclic Compounds. 2013;**49**(9):1289-1300

[13] Ilyin IY, Pushkarevsky NA, Shapovalov SS, Pasynskii AA, Konchenko SN, Scheer M, et al. Chalcogen arsenide clusters of iron with a functional carboxyl group: Synthesis, structures, and thermolysis. Russian Journal of Coordination Chemistry [Internet]. 2012;**38**(10):662-670. Available from: http://link.springer.com/10.1134/S1070328412100065

[14] Shieh M. Recent developments of tellurium- and selenium-containing iron carbonyl clusters. Journal of Cluster Science [Internet]. 1999;**10**(1):3-36. Available from: http://www.scopus.com/inward/record.url?scp=0033481657&partnerID=8YFLogxK%5Cn http://www.scopus.com/inward/citedby.url?scp=0033481657&partnerID=8YFL ogxK

[15] Du L, Wu Z, Shi Y, Li S, Jiang Y. Comparison of near-infrared absorption and photoresponse of silicon doped with Se and Te via fs-laser irradiation. Applied Physics B [Internet]. 2017;**123**(12):283. Available from: http://link.springer.com/10.1007/s00340- 017-6861-1

[16] Gupta A, Kumar S, Singh HB. Structural and reactivity aspects of organoselenium and tellurium cations. Proceedings of the National Academy of Sciences, India, Section A: Physical Sciences. 2016;**86**(4):465-498

[17] Király B, Tárkányi F, Takács S, Kovács Z. Excitation functions of proton induced nuclear reactions on natural tellurium up to 18 MeV for validation of isotopic cross sections. Journal of Radioanalytical and Nuclear Chemistry. 2006;**270**(2):369-378

[18] Zonaro E, Piacenza E, Presentato A, Monti F, Dell'Anna R, Lampis S, et al. *Ochrobactrum* sp. MPV1 from a dump of roasted pyrites can be exploited as bacterial catalyst for the biogenesis of selenium and tellurium nanoparticles. Microbial Cell Factories [Internet]. 2017;**16**(1):1-17. Available from: https://doi.org/10.1186/s12934-017-0826-2

[19] Yang F, Chen L, Hu Q, Pan G. Effect of the application of selenium on selenium content of soybean and its products. Biological Trace Element Research. 2003;**93**(1-3):249-256

[20] Wonorahardjo S, Wijaya AR, Suharti S. Surface behavior of Rhodamin and Tartrazine on silica-cellulose sol-gel surfaces by thin layer elution. The Journal of Pure and Applied Chemistry Research. 2016;**5**(May):48-54

[21] Wonorahardjo S, Ibnu MS, Budiasih E. Original research paper sulfur dioxide and ammonia gas reduction using coconut cellulose and acetylated. Cellulose. 2016;**17**(2):179-188

[22] Dasna IW, Golhen S, Ouahab L, Peña O, Guillevic J, Fettouhi M. 1-D mixed stack of coordinated and uncoordinated radicals in Mn II (NITpPy) 4 [N (CN) 2] 2 (NITpPy = nitronyl nitroxide radical). Journal of the Chemical Society, Dalton Transactions [Internet]. 2000;**2**:129-132. Available from: http://pubs.rsc.org/en/content/articlelanding/2000/dt/a908891a/unauth#!divAbstract

[23] Dasna IW, Golhen S, Ouahab L, Fettouhi M, Pena O, Daro N, et al. Synthesis, X-ray crystal structures and magnetic properties of Cu (II)(NITpPy)$_2$ [N(CN)$_2$]$_2$ solv (NITpPy = nitronyl nitroxide radical, solv= H_2O or CH_3CN). From discrete molecules to 2-D polymeric coordination compounds. Inorganica Chimica Acta [Internet]. 2001;**326**(1):37-46. Available from: https://www.sciencedirect.com/science/article/pii/S0020169301005928

[24] Fariati F, Istikfaroh N, Effendy E, Ayu Darojah L. Synthesis and characterization of coordination compounds of silver(I) nitrite with ligands ethylenethiourea and N,N'-diethylthiourea. The Journal of Pure and Applied Chemistry Research [Internet]. 2016; **5**(3):142-147. Available from: http://jpacr.ub.ac.id/index.php/jpacr/article/view/253/pdf

[25] Dasna I, Golhen S, Ouahab L, Daro N, Sutter J-P. Ferromagnetic interactions in Mn(II) coordination complex containing nitronyl nitroxide radical and silver–dicyanide anion: Structure and magnetic studies of Mn II (NITpPy) 2 [Ag (CN)$_2$]$_2$. Polyhedron [Internet]. 2001;**20**(11):1371-1374. Available from: https://www.sciencedirect.com/science/article/pii/S0277538701006210

[26] Dasna I, Golhen S, Ouahab L, Daro N, Sutter J-P. Synthesis, X-ray crystal structures and magnetic properties of Cu II and Mn II complexes containing imino nitroxide radicals and a dicyanamide anion. New Journal of Chemistry [Internet]. 2001;**25**(12): 1572-1576. Available from: http://pubs.rsc.org/en/content/articlelanding/2001/nj/b104773f/unauth#!divAbstract

[27] Knyukshto VN, Volkovich DI, Gladkov LL, Kuzmitsky VA, Ul-Haque A, Popkova IA, et al. Phenyl substituted Mg porphyrazines: The effect of annulation of a chalcogen-containing heterocycle on the spectral-luminescent properties. Optics and Spectroscopy [Internet]. 2012;**113**(4):359-375. Available from: http://link.springer.com/10.1134/S0030400X12070119

[28] Naumov NG, Brylev KA, Mironov YV, Cordier S, Fedorov VE. Octahedral clusters with mixed inner ligand environment: Self-assembly, modification and isomerism. Journal of Structural Chemistry [Internet]. 2014;**55**(8):1371-1389. Available from: http://link.springer.com/10.1134/S0022476614080010

[29] Naumov NG, Virovets AV, Fedorov VE, Branch S. Octahedral rhenium(ILL) chalcocyanide cluster anions: Synthesis, structure, and solid state design. Journal of Structural Chemistry. 2000;**41**(3)

[30] Fajaroh F, Setyawan H, Sutrisno N, Wonorahardjo S. To enhance the purity and crystallinity of magnetite nanoparticles prepared by surfactant-free electrochemical method by imposing higher voltage. AIP Conference Proceedings. 2014;**1586**:179-182

[31] Jang HJ, Lee HJ, Yoon DK, Ji DS, Kim JH, Lee CH. Antioxidant and antimicrobial activities of fresh garlic and aged garlic by-products extracted with different solvents. Food Science and Biotechnology [Internet]. 2018;**27**(1):219-225. Available from: https://doi.org/10.1007/s10068-017-0246-4

[32] Sun YE, Wang W. Changes in nutritional and bio-functional compounds and antioxidant capacity during black garlic processing. Journal of Food Science and Technology. 2018;**55**(2):479-488

[33] Beato VM, Orgaz F, Mansilla F, Montaño A. Changes in phenolic compounds in garlic (*Allium sativum* L.) owing to the cultivar and location of growth. Plant Foods for Human Nutrition. 2011;**66**(3):218-223

[34] Afzal M, Ali M, Thomson M, Armstrong D. Garlic and its medicinal potential. Inflammopharmacology. 2000;**8**(2):123-148

[35] Nam H, Jung H, Kim Y, Kim B, Kim KH, Park SJ, et al. Aged black garlic extract regulates lipid metabolism by inhibiting lipogenesis and promoting lipolysis in mature 3T3-L1 adipocytes. Food Science and Biotechnology [Internet]. 2017;**27**:3-7. Available from: http://link.springer.com/10.1007/s10068-017-0268-y

[36] Sato E, Kohno M, Hamano H, Niwano Y. Increased anti-oxidative potency of garlic by spontaneous short-term fermentation. Plant Foods for Human Nutrition. 2006; **61**(4):157-160

[37] Lanzotti V. Bioactive polar natural compounds from garlic and onions. Phytochemistry Reviews. 2012;**11**(2-3):179-196

[38] Da ZZ, Liang YZ, Xu CJ. Comparing chemical fingerprints of herbal medicines using modified window target-testing factor analysis. Analytical and Bioanalytical Chemistry. 2005;**381**(4):913-924

[39] Park EJ, Pezzuto JM. Botanicals in cancer chemoprevention. Cancer Metastasis Reviews. 2002;**21**(3-4):231-255

[40] Zhu Q, Kakino K, Nogami C, Ohnuki K, Shimizu K. An LC-MS/MS-SRM method for simultaneous quantification of four representative organosulfur compounds in garlic products. Food Analytical Methods [Internet]. 2016;**9**(12):3378-3384. Available from: http://dx.doi.org/10.1007/s12161-016-0535-1

[41] Zalepugin DY, Tilkunova NA, Chernyshova IV. Stability of thiosulfinates from garlic (*Allium sativum* L.) supercritical extracts in polar and nonpolar solvents. Russian Journal of Physical Chemistry B [Internet]. 2015;**9**(7):1032-1042. Available from: http://link.springer.com/10.1134/S1990793115070143

[42] Zalepugin DY, Til'kunova NA, Chernyshova IV, Mulyukin AL. Sulfur-containing components of supercritical garlic extracts and their synthetic analogs as potential biocides. Russian Journal of Physical Chemistry B [Internet]. 2013;**7**(7):843-848. Available from: http://link.springer.com/10.1134/S1990793113070154

[43] Yenigün B, Güvenilir Y. Partial purification and kinetic characterization of acid phosphatase from garlic seedling. Applied Biochemistry and Biotechnology: Part A: Enzyme Engineering and Biotechnology [Internet]. 2003;**107**(1-3):677-688. Available from: https://www.scopus.com/inward/record.uri?eid=2-s2.0-0038748235&doi=10.1385%2FABAB%3A107%3A1-3%3A677&partnerID=40&md5=8c453555f4a1d895ce46d7fa12de161e

[44] Harborne JB. Phytochemical methods; a guide to modern techniques of plant analysis. Journal of Chemical Information and Modeling. 1998;**3**:317

[45] Endo A, Imai Y, Nakamura M, Yanagisawa E, Taguchi T, Torii K, et al. Distinct intraspecific variations of garlic (*Allium sativum* L.) revealed by the exon-intron sequences of the alliinase gene. Journal of Natural Medicines. 2014;**68**(2):442-447

[46] Khaghani S, Ghanbari D. Magnetic and photo-catalyst Fe_3O_4–Ag nanocomposite: Green preparation of silver and magnetite nanoparticles by garlic extract. Journal of Materials Science: Materials in Electronics. 2017;**28**(3):2877-2886

[47] Saman N, Johari K, Mat H. Synthesis and characterization of sulfur-functionalized silica materials towards developing adsorbents for mercury removal from aqueous solutions. Microporous and Mesoporous Materials [Internet]. 2014;**194**:38-45. Available from: http://dx.doi.org/10.1016/j.micromeso.2014.03.036

[48] Tull BR, Winkler MT, Mazur E. The role of diffusion in broadband infrared absorption in chalcogen-doped silicon. Applied Physics A: Materials Science & Processing. 2009;**96**(2):327-334

[49] Romanov RI, Fominski VY, Shelyakov AV, Golubkov GV. Structure and catalytic properties of MoSex thin films containing Mo nanoparticles in electrochemical production of hydrogen in solution. Russian Journal of Physical Chemistry B [Internet]. 2016;**10**(2):238-244. Available from: http://link.springer.com/10.1134/S1990793116020238

[50] Sharma A, Mehta N. Optical characterization of tin containing novel chalcogen rich glassy semiconductors. Optical and Quantum Electronics [Internet]. 2018;**50**(2):1-13. Available from: https://doi.org/10.1007/s11082-018-1386-3

[51] Mahmudov KT, Kopylovich MN, Guedes da Silva MFC, Pombeiro AJL. Chalcogen bonding in synthesis, catalysis and design of materials. Dalton Transactions [Internet]. 2017;**46**(31):10121-10138. Available from: http://xlink.rsc.org/?DOI=C7DT01685A

[52] Shishkin OV, Omelchenko IV, Kalyuzhny AL, Paponov BV. Intramolecular S···O chalcogen bond in thioindirubin. Structural Chemistry. 2010;**21**(5):1005-1011

[53] Pascoe DJ, Ling KB, Cockroft SL. The origin of chalcogen-bonding interactions. Journal of the American Chemical Society. 2017;**139**(42):15160-15167. DOI: 10.1021/jacs.7b08511

[54] Pleshchev VG, Baranov NV, Mart'yanova IA. Structural phase transformations and physical properties of intercalation compounds in the $Cr_{0.5}Ti(Se_{1-x}Te_x)_2$ system. Physics of the Solid State [Internet]. 2006;**48**(10):1959-1964. Available from: http://link.springer.com/10.1134/S1063783406100246

[55] Baranov NV, Pleshchev VG, Sherokalova EM, Selezneva NV, Volegov AS. Influence of the chalcogen substitution on the character of magnetic ordering in $Fe_{0.5}TiS_{2-x}Se_x$ intercalated compounds. Physics of the Solid State [Internet]. 2011;**53**(4):701-706. Available from: http://link.springer.com/10.1134/S1063783411040044

[56] Posth O et. al. Classification of analysis methods for characterization of magnetic nanoparticle properties. In: Proceeding of XXI IMEKO World Congress, Measurement in Research and Industry. Prague, Czech Republic. August 30 September 4, 2015

[57] Badaloni M, Rossi M, Chiappini G, Lava P, Debruyne D. Impact of experimental uncertainties on the identification of mechanical material properties using DIC. Experimental Mechanics. 2015;**55**(8):1411-1426

[58] Bordovsky GA, Gladkikh PV, Kozhokar MY, Marchenko AV, Seregin PP, Terukov EI. Tin impurity centers in glassy germanium chalcogenides. Semiconductors [Internet]. 2011;**45**(10):1346-1351. Available from: http://link.springer.com/10.1134/S1063782611100058

[59] Bordovsky GA, Nemov SA, Marchenko AV, Seregin PP. Mössbauer studies of two- electron centers with negative correlation energy in crystalline and amorphous semiconductors. Semiconductors [Internet]. 2012;**46**(1):1-21. Available from: http://link.springer.com/10.1134/S1063782612010071

[60] Ajibade PA, Zulu NH, Oyedeji AO. Synthesis, characterization, and antibacterial studies of some metal complexes of dialkyl thiourea: The X-ray single crystal structure of $[CoCl_2(detu)_2]$. Synthesis and Reactivity in Inorganic Metal-Organic and Nano-Metal Chemistry [Internet]. 2013;**43**(5):524-531. Available from: http://www.tandfonline.com/doi/abs/10.1080/15533174.2012.741179

[61] Ajibade PA, Zulu NH. Synthesis, characterization, and antibacterial activity of metal complexes of phenylthiourea: The X-ray single crystal structure of $[Zn(SC(NH_2)NHC_6H_5)_2(OOCCH_3)_2]\,C_2HOH$. Journal of Coordination Chemistry. 2010;**63**(18):3229-3239

[62] Preetha KC, Remadevi TL. Band gap engineering in PbSe thin films from near-infrared to visible region by photochemical deposition method. Journal of Materials Science: Materials in Electronics. 2014;**25**(4):1783-1791

[63] Nucci A, Marino-Merlo F, De Nisco M, Pedatella S, Rossi F, Jacob C, et al. Se-(2-aminoalkyl)selenocysteines as biochemical redox agents. A tool to contrast cell injury induced by aflatoxin B1 in HepG2 cells. Amino Acids. 2014;**46**(2):459-470

[64] Jain VK. Synthesis and characterization of single-source molecular precursors for the preparation of metal chalcogenides. Journal of Chemical Sciences. 2006;**118**(6):547-552

[65] Malik MR, Vasylyeva V, Merz K, Metzler-Nolte N, Saleem M, Ali S, et al. Synthesis, crystal structures, antimicrobial properties and enzyme inhibition studies of zinc(II) complexes of thiones. Inorganica Chimica Acta [Internet]. 2011;**376**(1):207-211. Available from: http://dx.doi.org/10.1016/j.ica.2011.06.017

[66] Del CR, Criado JJ, Gheorghe R, González FJ, Hermosa MR, Sanz F, et al. N- benzoyl-N′-alkylthioureas and their complexes with Ni(II), Co(III) and Pt(II)—Crystal structure of 3-benzoyl-1-butyl-1-methyl-thiourea: Activity against fungi and yeast. Journal of Inorganic Biochemistry. 2004;**98**(8):1307-1314

[67] Rooney M, Roberts JC, Murray GM, Romenesko BM. Advanced materials: Challenges and opportunities. 2000;**21**(4):516-527

[68] Wonorahardjo S, Ibnu MS, Budiasih E. Sulfur dioxide and ammonia gas reduction using coconut cellulose and acetylated cellulose. Scientific Study and Research: Chemistry and Chemical Engineering, Biotechnology, Food Industry. 2016;**17**(2):179-188

[69] Kozytskiy AV, Stroyuk OL, Raevskaya AE, Kuchmy SY. Photoelectrochemical solar cells with semiconductor nanoparticles and liquid electrolytes: A review. Theoretical and Experimental Chemistry. 2017;**53**(3):145-179

[70] Dutt MB, Sharma BL. 3 Diffusion in compound semiconductors. Diffusion in Semiconductors SE- 10 [Internet]. 1998;**33A**:1-63. Available from: http://dx.doi.org/10.1007/10426818_10

[71] Stuart BH. Infrared spectroscopy: Fundamentals and applications [Internet]. Methods. 2004;**8**:224. Available from: http://doi.wiley.com/10.1002/0470011149

[72] Frumar M et al. Crystalline and amorphous chalcogenides, high-tech materials with structural disorder and many important applications. In: Bardosova M, Wagner T, editors. Nanomaterials and Nanoarchitectures: A Complex Review of Current Hot Topics and their Applications. Netherlands: Springer; 2015

[73] Zonkhoeva EL, Sanzhanova SS. Infrared spectroscopy study of the sorption of selenium(IV) on natural zeolites. Russian Journal of Physical Chemistry A [Internet]. 2011;**85**(7): 1233-1236. Available from: http://link.springer.com/10.1134/S0036024411070399

[74] Bouroushian M. In: Scholz F, editor. Electrochemistry of Metal Chalcogenides. Heidelberg: Springer Berlin Heidelberg; 2010

[75] Baranov NV, Gerasimov EG, Mushnikov NV. Magnetism of compounds with a layered crystal structure. Physics of Metals and Metallography [Internet]. 2011;**112**(7):711-744. Available from: http://dx.doi.org/10.1134/S0031918X11070039

[76] Nilges T, Bawohl M, Lange S, Messel J, Osters O. Highly dynamic chalcogen chains in silver(I) (Poly)chalcogenide halides: A new concept for thermoelectrics? Journal of Electronic Materials. 2010;**39**(9):2096-2104

[77] Kehr M, Fatkullin N, Kimmich R. Molecular diffusion on a time scale between nano- and milliseconds probed by field-cycling NMR relaxometry of intermolecular dipolar interactions: Application to polymer melts. The Journal of Chemical Physics. 2007;**126**(9): 094903-094903-8

[78] Fatkullin N, Kimmich R. Theory of field-gradient NMR diffusometry of polymer segment displacements in the tube-reptation model. Physical Review E [Internet]. 1995; **52**(3):3273-3276. Available from: http://link.aps.org/doi/10.1103/PhysRevE.52.3273

[79] Leszczynski J, (series editor). Challenges and advances in computational chemistry and physics. In: Scheiner S, editor. Noncovalent Forces. Cham: Springer; 2006 [Internet]. 1-680 p. Available from: http://link.springer.com/content/pdf/10.1007/1-4020-4850-5.pdf%5Cnpapers2://publication/uuid/F8F76DC1-C945-42AF-AD3E-05DEC2D420DE

[80] Saito M, Kikuchi K. Infrared optical fiber sensors. Optical Review. 1997;**4**(5):527-538

[81] Mattea C, Kimmich R, Ardelean I, Wonorahardjo S, Farrher G. Molecular exchange dynamics in partially filled microscale and nanoscale pores of silica glasses studied by field-cycling nuclear magnetic resonance relaxometry. The Journal of Chemical Physics. 2004;**121**(21):10648-10656

[82] Ardelean I, Mattea C, Farrher G, Wonorahardjo S, Kimmich R. Nuclear magnetic resonance study of the vapor phase contribution to diffusion in nanoporous glasses partially filled with water and cyclohexane. The Journal of Chemical Physics [Internet]. 2003;**119**(19):10358. Available from: http://link.aip.org/link/JCPSA6/v119/i19/p10358/s1&Agg=doi

[83] Kimmich R. NMR-Tomography, Diffusometry, Relaxometry. Berlin: Springer Berlin Heidelberg; 1997. 526p

[84] Wonorahardjo S, Ball G, Hook J. Moran G. NMR relaxation studies of porous sol-gel glasses. Magnetic Resonance Imaging [Internet]. 1998;**16**(5-6):511-513. Available from: http://www.ncbi.nlm.nih.gov/pubmed/9803899

[85] Wonorahardjo S, Ball GE, Hook J, Moran G. 2H NMR relaxation monitoring of gelation in tetramethoxysilane sol-gels. Journal of Non-Crystalline Solids [Internet]. 2000;**271** (1–2):137-146. Available from: http://linkinghub.elsevier.com/retrieve/pii/S0022309300000685

[86] Nurindah N, Wonorahardjo S, Adi Sunarto D, Sujak S. Chemical cues in tritrophic interaction on biocontrol of insect pest. The Journal of Pure and Applied Chemistry Research [Internet]. 2017;**6**(1):49-56. Available from: http://jpacr.ub.ac.id/index.php/jpacr/article/view/282/pdf

[87] Sjöström J. The discourse of chemistry (and beyond). HYLE International Journal for Phylosophy of Chemistry. 2007;**13**(2):83-97

[88] Sjöström J, Talanquer V. Humanizing chemistry education: From simple contextualization to multifaceted problematization. Journal of Chemical Education. 2014;**91**(8):1125-1131

[89] Sjöström J, Eilks I, Zuin VG. Towards eco-reflexive science education. Science & Education [Internet]. 2016;**25**(3-4):321-341. Available from: http://link.springer.com/10.1007/s11191-016-9818-6

Chapter 3

The Characterization of a Newly Layered Bimetallic Hydrogen Selenite Copper-Selenium: Synthesis and Structure

Mohamed Loukil

Additional information is available at the end of the chapter

http://dx.doi.org/10.5772/intechopen.76310

Abstract

Single-crystal X-ray diffraction data were used to solve the structure of a newly layered copper-selenium hydrogen selenite and further refined to a final reliability factor, R_1 = 0.038. This structure was found to have an orthorhombic space group $PBn2_1$, with a = 7.1753(4) Å, b = 9.0743(4) Å, c = 17.725(9) Å, V = 1154.06(10) $Å^3$, and Z = 4. Although this structure may be described to exhibit a bidimensional structure, it is actually three-dimensional in shape. The bidimensional structure is made up of layers, parallel to the (010) plane, which contain copper atoms and $(HSeO_3)^-$ anions with sheets interconnected by $[CuCl_3(H_2O)_3]$ groups. Bond valence sum calculations were used to evaluate the Se and Cu oxidation states. Both the infrared (IR) and Raman spectra were obtained and employed to confirm the presence of hydrogen selenites (Se—O—H). Also, the dielectric constant at different frequencies and temperatures revealed a phase transition at 383 K.

Keywords: infrared, structural study, dielectric properties, bimetallic hydrogen selenites

1. Introduction

During the past years, lot of interest has been shown in hydrogen selenite chemistry motivating research focused on expanding the knowledge of the structural and bonding principles of this ligand. There is an important number of divalent metal hydrogen selenite crystal structures reported in the literature. For example, $M(HSeO_3)_2$ (where M: Cu, Mg, Sr, Ba) [1, 2]; $M(HSeO_3)_2{\cdot}H_2O$ (where M: Ca, Cu) [3–6]; $M(HSeO_3)_2{\cdot}NH_4Cl$ (where M: Cu) [7]; $[Cu(HSeO_3)_2Cu_xM_{1-x}Cl_2(H_2O)_4]$ with M = Cu, Co, Mn, Ni, and Zn [8]. This family has a three-dimensional structure, but it may be considered as being derived from the $[Cu(HSeO_3)_2]$-type structure.

Compounds exhibiting mixed valences (like Se^{4+} and Cu^{2+}) are at the center of many studies owing to their potential applications in relation to the electronic exchange. For almost all of these compounds, except $[Cu(HSeO_3)_2]$, magnetic measurements have revealed the occurrence of weak ferromagnetism at low temperature (T ~ 10–20 K) for which a tentative explanation is offered for this peculiar property in agreement with other authors [8, 9]. In this study, the crystal structure of the compound $[Cu_{0.335}Se_{0.582}(HSeO_3)_2CuCl_3(H_2O)_3]$, herein presented, was obtained using an X-ray single structure and various spectroscopic (IR and Raman) characterization, as well as dielectric measurements.

2. Materials and methods

The experiments were carried out using a single crystal of $[Cu_{0.335}Se_{0.582}(HSeO_3)_2CuCl_3(H_2O)_3]$ grown by slow evaporation from a mixture of hydrochloric acid containing stoichiometric $CuCl_2$-SeO_2 at room temperature in the ratio 1/2. Blue thin rectangular parallelepiped crystals were grown after vaporizing in air for 15 days approximately. The determination of $[Cu_{0.335}Se_{0.582}(HSeO_3)_2CuCl_3(H_2O)_3]$ formula was achieved by the crystal structure refinement approach at room temperature.

For electrical impedance measurements (in the range 1–10 KHz), a Hewlett-Packard 4192 ALF automatic bridge monitored by a HP Vectra microcomputer was used. Prepared dense translucent pellets with approximately a diameter of 8 mm and a thickness of 1–1.2 mm, covered with graphite electrodes were utilized for the measurements.

The measurement of electrical impedances were equally carried out in the range, 1–10 KHz, using a Hewlett-Packard 4192 ALF automatic bridge monitored by a HP Vectra microcomputer. For these types of experiments, dense translucent pellet samples were prepared, with a diameter of 8 mm and thickness between 1 and 1.2 mm. All the pellets were then covered with graphite electrodes prior to measurements.

A modern nondispersive Fourier Transform (FT-IR) spectrometer (Perkin-Elmer 1750 spectrophotometer IR-470) was employed for the characterization of the crystalline powders after mixing with KBr, with very notable IR-active functional groups found in the samples investigated. Scans of IR spectra were recorded in in the range 400–4000 cm^{-1} without apodization. To record Raman spectra of the solid samples for the study of the various phases, a conventional scanning Raman instrument (Horiba Jobin Yvan HR800 microcomputer system), equipped with a Spex 1403 double monochromator (with a pair of 600 grooves/mm gratings) and a Hamamatsu 928 photomultiplier detector was used. Solid materials were sampled at different temperatures for this analysis. During the recording of prominent Raman peaks, excitation radiation from the instrument was fulfilled by a coherent radiation emitted by a He-Neon laser operating at a wavelength of 633 nm, with an output laser power of 50 mW. In order to acquire high-resolution Raman spectra, the spectral resolution of the slit width varied from 3 to 1 cm^{-1}.

To study the crystal structure of $[Cu_{0.335}Se_{0.582}(HSeO_3)_2CuCl_3(H_2O)_3]$, an APEX II diffractometer (powder XRD) fitted with graphite-crystal monochromated Mo Kα radiation (0.71073 Å) was employed. In this study, a total of 3093 reflections were collected, among which only 2803 reflections, namely those for which $I > 2\sigma$ (I), were actually used in the determination and refinement of the structure. Corrections were made for Lorentz-Polarization and absorption effects. **Table 1** presents the data collection procedure and structure refinement at room temperature.

A three-dimensional Patterson synthesis approach was used to determine the selenium atoms positions in the compound. On the one hand, the Fourier function allowed for the localization of the chlorine (Cl), copper (Cu), and oxygen (O) atoms. On the other hand, the hydrogen atoms were localized from a difference Fourier synthesis and introduced as fixed contributors. Conversely, all the non-hydrogen atoms were typically assigned anisotropic thermal displacements. The structure solution and refinement were carried out using SHELX programs [10, 11]. The bond lengths and angles are given in **Table 2**.

Crystallographic data	**T = 296(2) K**
Formula	$[Cu_{0.335}Se_{0.582}(HSeO_3)_2CuCl_3(H_2O)_3]$
Formula weight	547.15
Space group	$Pbn2_1$
a (Å)	7.1753(4)
b (Å)	9.0743(4)
c (Å)	17.7246(9)
V (Å^3)	1154.06(10)
Z	4
ρ_{calc} (g/cm^3)	3.149
μ (mm^{-1})	11.359
Crystal size (mm^3)	$0.05 \times 0.04 \times 0.03$
Crystal shape	Octahedral
F(000)	1030
Data collection instrument	Kappa-APEX II
Radiation, graphite	λMo K_α (0.71073 Å)
θ range for data collection (°)	6.80–30.57
Index ranges	$-10 \le h \le 9; -12 \le k \le 12; -23 \le l \le 25$
Total reflections	8605
Reflection with ($F > 4\sigma(F)$)	3093
R(F) (%)	3.38
W_{R2} (%)	1.74

Table 1. Crystal structure data and experimental conditions for the structure determination of $[Cu_{0.335}Se_{0.582}(HSeO_3)_2CuCl_3(H_2O)_3]$.

a: SeO_3 polyhedron	
Se1—O4 = 1.70(5)	O5—Se1—O7 = 102(2)
Se1—O5 = 1.70(5)	O5—Se1—O4 = 102(2)
Se1—O7 = 1.71(5)	O7—Se1—O4 = 98(2)
Se3—O3 = 1.67(4)	O3—Se3—O6 = 98(3)
Se3—O6 = 1.66(6)	O3—Se3—O8 = 101(3)
Se3—O8 = 1.77(5)	O6—Se3—O8 = 100(2)
b: $Cu(Se)O_4Cl_2$	
Cu2/Se2—O3 = 1.98(4)	O4(b)—Cu2/Se2—O6(a) = 91(2)
Cu2(c)/Se2(c)—O4(b) = 1.91(4)	O4(b)—Cu2/Se2—O3 = 179(3)
Cu2/Se2—O5 = 1.94(4)	O6(a)—Cu2/Se2—O3 = 90(2)
Cu2(d)/Se2(d)—O6(a) = 2.01(5)	O4(b)—Cu2/Se2—O5 = 89(2)
Cu2/Se2—Cl2 = 2.77(2)	O6(a)—Cu2/Se2—O5 = 180(4)
Cu2/Se2—Cl3(e) = 2.80(2)	O3—Cu2/Se2—O5 = 90.1(2)
	Cl2—Cu2/Se2—O6(a) = 89.52(8)
	O3—Cu2/Se2—Cl3(e) = 89.11(5)
	Cl2—Cu2/Se2—O3 = 91.1(4)
	Cl2—Cu2/Se2—O4(b) = 89.83(3)
	O5—Cu2/Se2—Cl2 = 94.18(7)
c: $CuCl_3(H_2O)$ octahedron	
Cu1—Cl1 = 2.66 (8)	O1—Cu1—O2 = 88.1(3)
Cu1—Cl2 = 2.30 (2)	O1—Cu1—Cl3 = 176(2)
Cu1—Cl3 = 2.30(2)	O1—Cu1—Cl2 = 90.9(2)
Cu1—O1 = 1.99(7)	O2—Cu1—Cl2 = 176.1(5)
Cu1—O2 = 2.02(6)	O2—Cu1—Cl1 = 83(2)
Cu1—O9 = 2.21(3)	Cl3—Cu1—Cl2 = 92.2(3)
	Cl3—Cu1—Cl1 = 94.1(3)

Symmetry code: a: $-x + 1/2, y + 1/2, z$; b: $-x + 3/2, y + 1/2, z$; c: $-x + 3/2, y - 1/2, z$; d: $-x + 1/2, y - 1/2, z$; e: $-x + 1, -y + 2, z + 1/2$.

Table 2. Interatomic distances for $[Cu_{0.335}Se_{0.582}(HSeO_3)_2CuCl_3(H_2O)_3]$ samples (this study).

3. Results and discussion

3.1. Structure description

From the charge balance in $[Cu_{0.335}Se_{0.582}(HSeO_3)_2CuCl_3(H_2O)_3]$, it can be suggested that the average oxidation state of Cu(2)/Se(2) is equal to 3, which would fit to 33.5% of Cu^{2+} and 58.2% of Se^{4+}. This outcome was confirmed by performing a calculation of bond valence sums around the centers of the cation sites. The steps and expressions used in the calculation of the bond valence are published in [12]. More specifically, the bond valence (S_{ij}) is expressed as $S_{ij} = \exp[(R_0 - R_{ij})/B]$, where R_0 and B are the experimentally determined parameters and R_{ij} is the bond length of the cation-anion pair [12]. The sum of the bond valence ($\Sigma_j S_{ij}$) around an ion calculated must be equal to the formal valence (V_i) of this ion, based on the valence sum rule.

In this work, our calculation shows that, for the pyramidal sites (Se1 and Se3), the sum of bond valence is ~4, which is equal to selenium formal valence. Thus, the bond valence sums around the octahedral site of Cu(2) are typically consistent with the value +2.7, confirming the presence of selenium Se^{4+} and cuprite Cu^{2+} in the same site. It is also observed that the blue single crystal $[Cu_{0.335}Se_{0.582}(HSeO_3)_2CuCl_3(H_2O)_3]$ crystallizes in the orthorhombic system, space group $Pbn2_1$. Structurally, the crystal structure of $[Cu_{0.332}Se_{0.582}(HSeO_3)_2CuCl_3(H_2O)_3]$ represents a new type of structure for complexes of hydrogen selenites (**Figure 1**). The building blocks $[Cu_{0.335}Se_{0.582}(HSeO_3)_2]$ and $[CuCl_3(H_2O)_3]$, hereunder drawn, are arranged to form layers in the structure parallel to the (001) plane between which the lone pairs E are located (**Figure 2**). Due to the stereochemical activity of the lone pairs E, Se has very asymmetric coordination polyhedral SeO_3 pyramids.

Spatially, the high anisotropic distribution of anions observed around each cation is characteristically of a strong stereochemical activity of their electron lone pair E for the Se1 and Se3 atoms. The consequence for the coordination polyhedral is the description of a distorted SeO_3 triangular pyramid, in which the Se—O(7) and Se—O(8) are marginally longer than the other Se—O bonds (**Table 2**). Thereof, the lone pair E so directed to constitute the fourth vertex of an SeO_3E tetrahedron (**Figure 3**).

The O—Se—O with angles of values 98(2) and 102(2) formed from Se—O chemical bonds are situated on one side of the Se atom, whereas the other side is hitherto a "dead" zone around

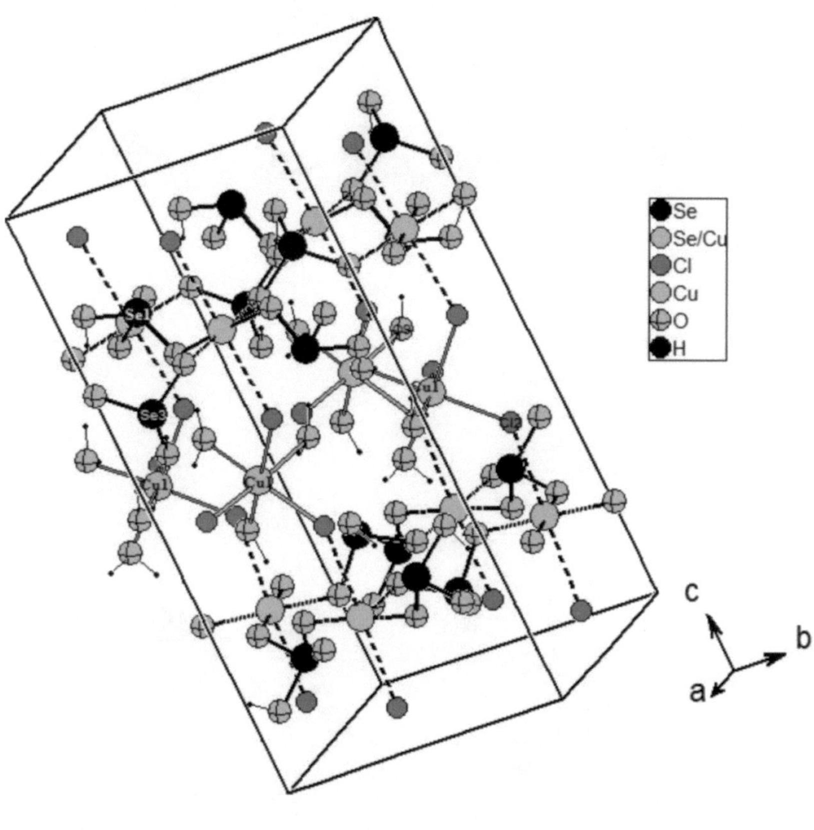

Figure 1. Perspective view of the $[Cu_{0.335}Se_{0.582}(HSeO_3)_2CuCl_3(H_2O)_3]$ unit cell content.

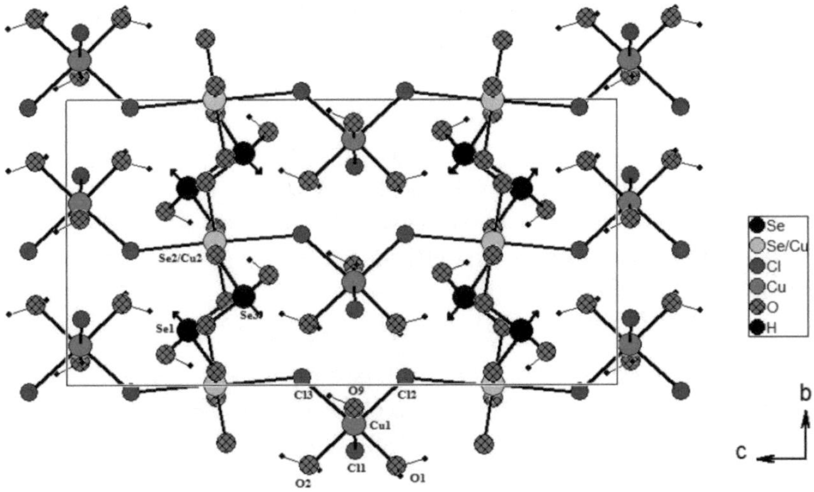

Figure 2. A projected along a-axis view of the $[Cu_{0.335}Se_{0.582}(HSeO_3)_2CuCl_3(H_2O)_3]$ unit cell content.

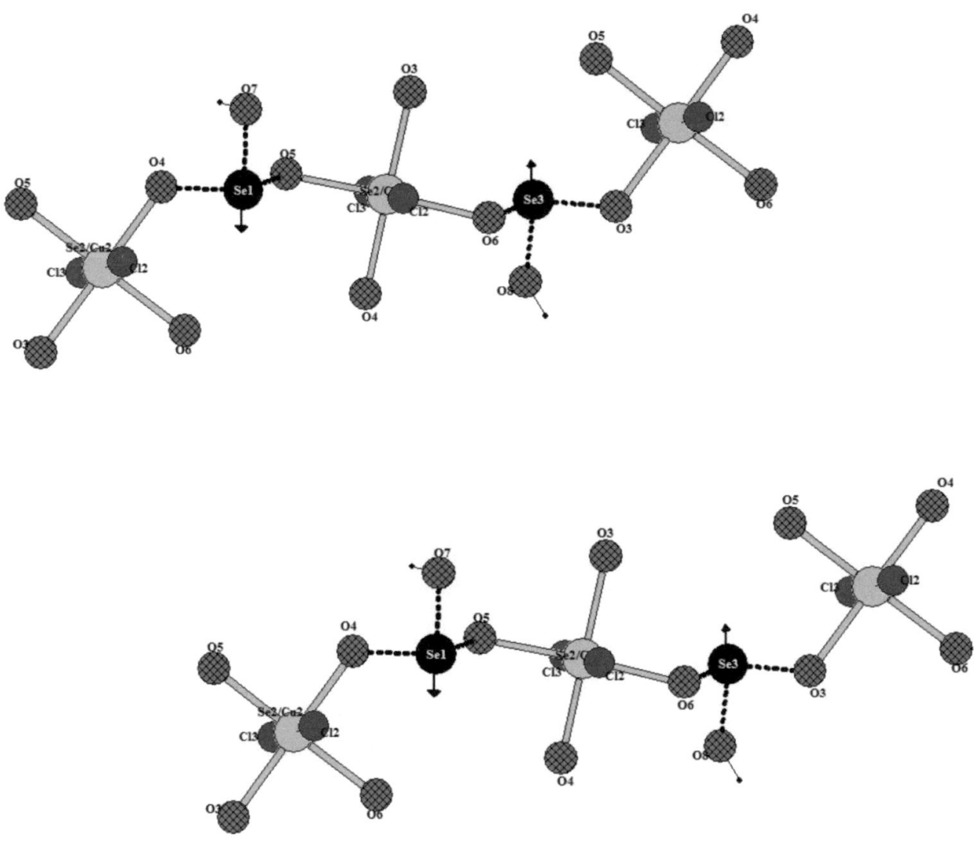

Figure 3. Environment of selenium Se(1), Se(3) cations.

the lone pair E of the Se atom. As such, the SeO_3 polyhedral has strong dipole moments, due to such a sharp asymmetry of the atomic arrangement of the first coordination spheres, which provide considerable dipole-dipole contribution to the inter-anion potentials. Taking a closer look at the structure in **Figure 4**, one can easily see that each Cu(1) atom is surrounded by three oxygen (O) atoms and three chlorine (Cl) atoms. This basically forms a slightly distorted octahedron with distances ranging from 1.99(7) to 2.21(3) Å for Cu—O and 2.30 (2) Å for Cu—Cl.

A remarkable deviation from full occupancy was exhibited in the occupancy of the Cu(2) site during refinement. This is an indication of a substitution with Se, resulting in final occupancies constrained in sum to 1.0, and refined to 0.335(4) and 0.582(2), respectively. It should be noted that the deviation from 1.0 is due to the mixed valence between Cu and Se, for Cu(2) and Se(2), respectively. As shown in **Figure 5**, the Cu(2)/Se(2) atoms are surrounded by four oxygen atoms and two chlorine atoms to form an irregular octahedron.

From the earlier arguments, the structure depicted in **Figure 2** can be ascribed to being formed by Cu(1), Se(2)/Cu(2) polyhedral that structurally shares chlorine (Cl) corners in infinite chains along the direction [001]. Therein, the sequential metal atoms in the chain trend schematically following Cu(1)—Se(2)/Cu(2)—Cu(1)′—Se(2)/Cu(2)—Cu(1), whereby the prime refers to the polyhedral generated by the symmetry operation: $1 - x$, $2 - y$, $-0.5 + z$, in the [010] direction. As a result, the three-dimensional network is thus formed. It can be argued that the lattice cohesion may be strengthened by the hydrogen bonds within the layer (O(8)—H(7)—O(4)) and (O(7)—H(8)—O(3)) or outside the layer.

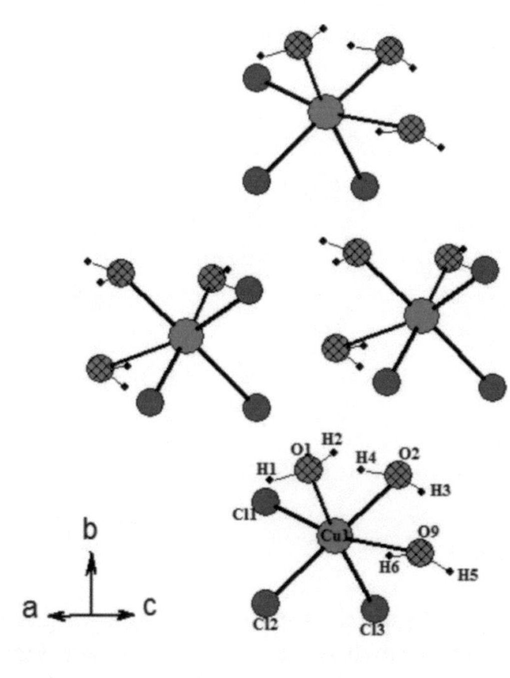

Figure 4. Structural environment of copper cations Cu(1).

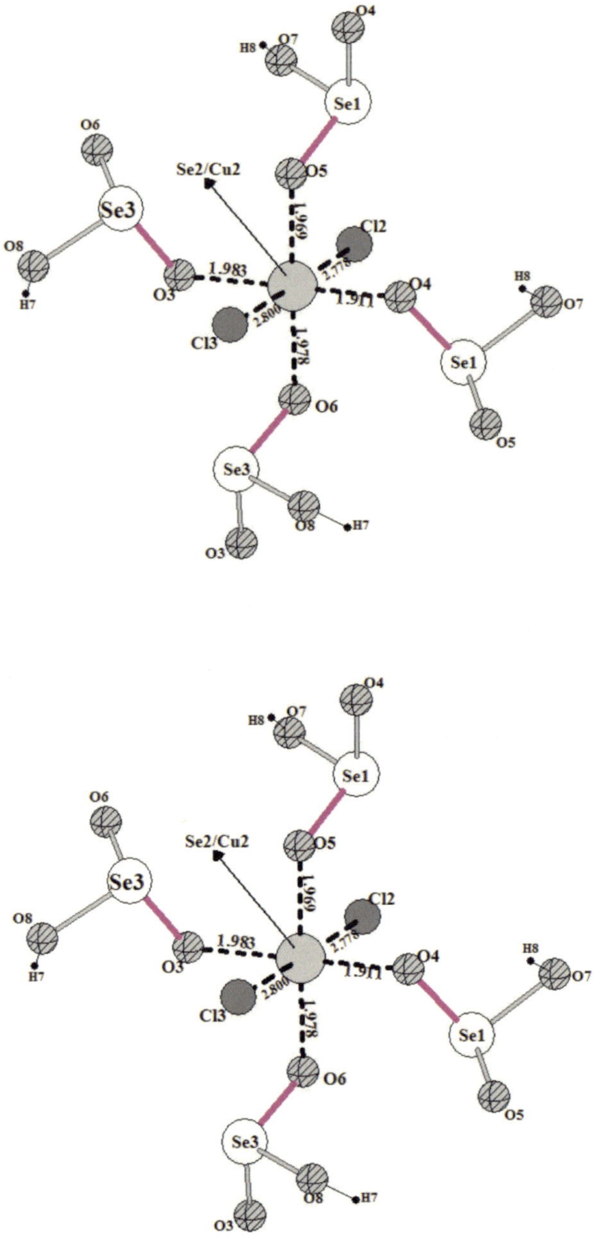

Figure 5. Environment of copper/selenium cations Cu(2)/Se(2).

3.2. Spectroscopic studies

In order to confirm the crystallographic results of the following compound: $[Cu_{0.335}Se_{0.582}(HSeO_3)_2CuCl_3(H_2O)_3]$, IR, and Raman spectroscopy were used. **Figure 6** shows that the IR spectrum is restricted to the mid-infrared frequency range: 400–4000 cm^{-1}.

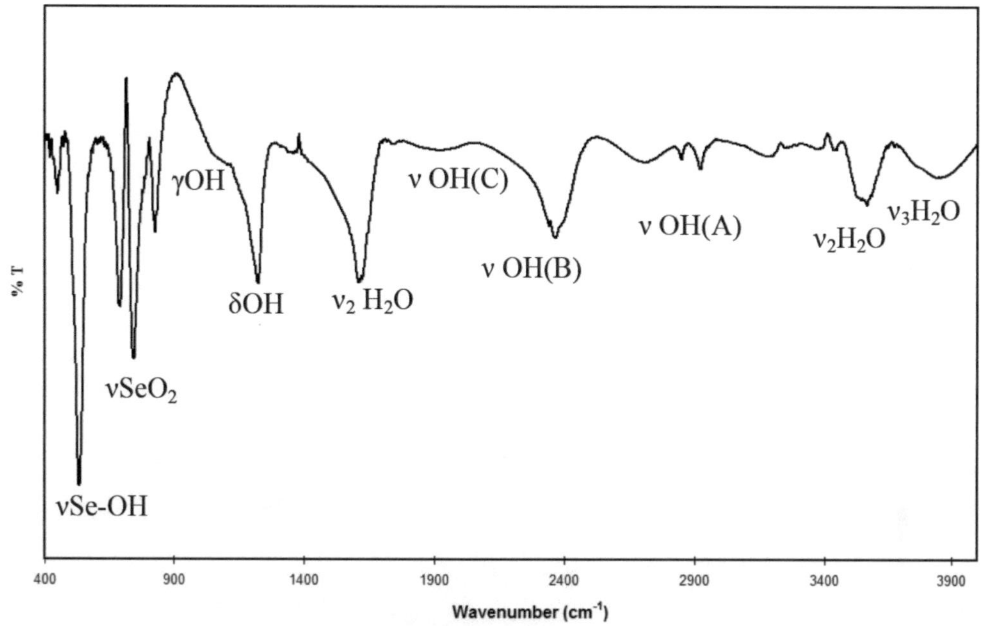

Figure 6. Infrared spectrum of $[Cu_{0.332}Se_{0.582}(HSeO_3)_2CuCl_3(H_2O)_3]$ at room temperature.

In this chapter, the band corresponding to the symmetric stretching vibrations of SeO_2 groups was observed at around 825 cm^{-1} in the Raman spectrum (**Figure 7**). Similarly, a strong intense broad band is observed in the infrared (IR) spectrum for this mode. These findings are in agreement with those reported by Cody and al. and Micka et al. for vibrational analysis on a series of alkali hydrogen selenites. From the work of these authors, the symmetric stretching vibrations are around 850 cm^{-1} [13–15].

From **Figure 7**, a band of very weak intensity is observed at 710 cm^{-1}, accompanied by a shoulder at 738 cm^{-1}, which is ascribed to asymmetric stretching vibrations of SeO_2 groups. In the 686–740 cm^{-1} region, a corresponding IR spectrum with an intense (broad) frequency absorption is present. In the literature [16, 17], these modes have been observed at a very much lower wave numbers than those in alkali hydrogen selenites.

Another observation is that copper (selenium) atoms are located at the center of CuO_4Cl_2 coordination octahedra. The axial Cu—Cl bonds are longer than the others, and they are coordinated to water molecules. It can be stated that the high spin d-configuration of Cu leads to Jahn-Teller distortions by Jahn and Teller [18] hitherto producing planar Cu—O bonds in the 1.91–1.98 Å range and an axial Cu—Cl distance of 2.77–2.80 Å. It can be ascertained that the planar oxygen (O) atoms are shared by Cu and Se atoms which can lead to the observed reduction in the stretching frequencies of the SeO_2 groups.

Typically, the stretching vibrations of the $HSeO_3^-$ ion (νSe—OH), which often characteristically appear in the 600–650 cm^{-1} region [16, 17], are also appearing at lower wave numbers (~627 cm^{-1}). In the IR spectrum, this mode is observed as an intense broad band at 532 cm^{-1}.

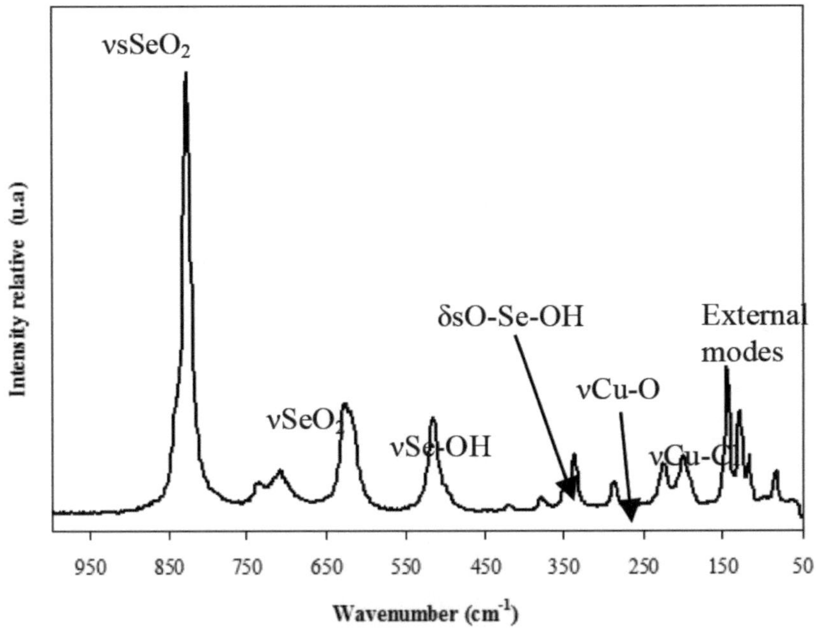

Figure 7. Raman spectrum of the compound at room temperature.

The O—O distance involving the Se—OH system with one of the equatorial oxygen atoms of the neighboring Cu(2)/Se(2)O_4Cl_2 group is 2.662 Å. The observed lowering of the Se—O(H) vibrations from the free-state values is a confirmation of the corresponding strong hydrogen bonds determined in the X-ray diffraction data. As presented in **Table 3**, it is seen that symmetric deformation vibrations of the $HSeO_3^-$ ion have given only weak bands of the Raman spectrum, while a medium intense broad band is obtained in the IR. In the corresponding asymmetric bending vibration, Raman spectrum shows medium intense bands with a weak band in the IR. A reduction in the symmetry of the $HSeO_3$ ion may be causing the changes in the activity of these modes. From the observed strong hydrogen bonding and the distortion of Cu(Se)O_4Cl_2 octahedra, we can deduce that the Jahn-Teller distortion affects the $HSeO_3$ vibrations.

Fundamentally, the hydrogen-bonded OH groups may lead to three vibrations, namely: ν(OH) stretching, the in-plane (OH), and the out-of-plane (OH) deformation vibrations. In fact, the stretching bands of strongly H-bonded systems are intense and usually built up of a number of unresolved components owing to strong interaction between the proton vibration and the ν(O...O) vibrations [19, 20]. It is also clearly elucidated from the literature [21, 22] that the broad ν(OH) band in Fermi resonance with the overtones of the ν(OH) modes splits into three bands A, B, and C. The A mode is typically observed as a strong broad band at 2850 cm^{-1} in the IR spectrum, while the B mode is obtained as a medium intense band at 2360 cm^{-1}. A medium intense broad band at 1900 cm^{-1} and a weak one at 1730 cm^{-1} are generally being assigned to C bands. From a practical viewpoint, the appearance of these bands confirms the existence of strong hydrogen bonds in the crystal. The in-plane bending ν(OH) vibrations are

Raman	IR	Assignments
	3553 mbr 3170 wbr	ν_1H_2O
	2920 w 2850 w 2694 wbr	νOH(A)
	2360 mbr	νOH(B)
	1900 vwbr 1730 vw	νOH(C)
	1606 mbr	ν_2H_2O
	1222 m	δOH
	1043–920	γOH
	896 m	
825 vs	825 s	$\nu sSeO_2$
738 wsh 710wbr 627 m	740 s 686 s	
515 m 501 sh	532 s	νSe—OH
425 vw	442 w	δSeO_2
380 vw		δasO—Se—OH
350 sh 337 m		δsO—Se—OH
288 w		Cu—O stretching
225 vw		stretching modes of Cu—Cl
201 mbr 143wbr 129 m 120w 83 m		External modes

Relative intensities: sh, shoulder; m, medium; w, weak; vw; very weak; mbr, medium broad; s, strong; vs., very strong.

Table 3. Assignment and frequencies (cm^{-1}) observed for IR and Raman spectra of [Cu0.335Se0.582($HSeO_3$) $2CuCl_3(H_2O)_3$] at room temperature.

less sensitive to the hydrogen bond strength than the ν(OH) mode [15]. A medium band observed at 1280 cm^{-1} in the IR spectrum is attributed to the in-plane ν(OH) bending mode and a medium intense broad band in the 870–940 cm^{-1} region to the ν(OH) mode.

Two prominent broad bands were observed in the stretching region of the typical water in the Raman spectra of the main (title) compound. Similarly, in the IR spectrum, a corresponding strong broad band with two distinct peaks exhibited at 3553 and 3170 cm^{-1} are noticeable for this mode. The bending mode of H_2O that appears at around 1606 cm^{-1} in the IR is noteworthy. The considerable shifting of stretching and bending frequencies from those of a free water molecule (H_2O) [23] may be an indication of the presence of strong hydrogen bonding in the new crystal. The external modes of the $HSeO_3$ ion, lattice modes of water, and metal-oxygen stretching modes appear approximately below 200 cm^{-1} [1].

3.3. Dielectric studies

Figure 8 shows an illustration of the temperature dependence of the dielectric constant (ε′) in the frequency range [1–10] KHz, and in the temperature region of 300–500 K obtained for $[Cu_{0.335}Se_{0.582}(HSeO_3)_2CuCl_3(H_2O)_3]$. These curves (**Figure 8**) exhibit the following characteristics:

(1) There is one anomaly in the dielectric constant ε′ observed at about 383 K, (2) there is a maxima in the permittivity curves, displaced to higher temperatures with increasing frequency, and (3) apparently, this is a transition which can be attributed to the "order-disorder" phase transition, probably characterizing the motion of H^+ diffusion related to the motion of $HSeO_3$ groups, as reported in the literature [16, 17].

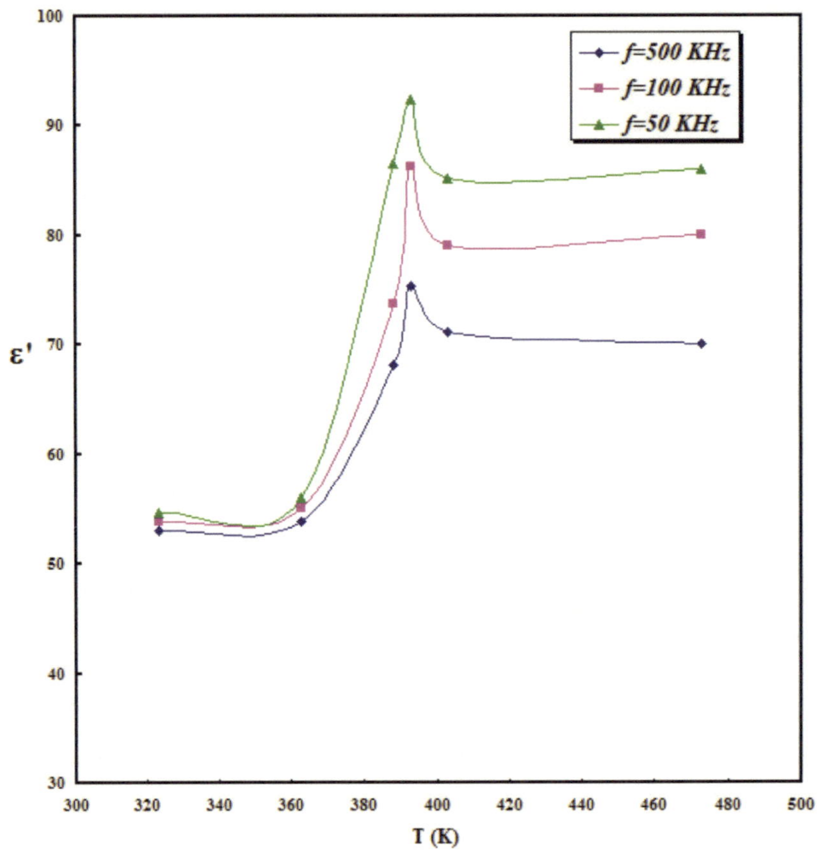

Figure 8. Temperature dependence of ε′ as a function of frequency.

4. Conclusion

In conclusion, the author ascertains that a novel substituted hydrogen selenites $[Cu_{0.335}Se_{0.582}(HSeO_3)_2CuCl_3(H_2O)_3]$, have been successfully prepared via slow evaporation method. The crystal structure of the novel compound is characterized by the presence of structural blocs with structures as such $[Cu_{0.335}Se_{0.582}(HSeO_3)_2]$ and $[CuCl_3(H_2O)_3]$. The principal compound is arranged to form layers in the structure parallel to the (001) plane between which the lone pairs E are located. So, the main feature of the structure of this compound is based on different coordination polyhedral, SeO_3 pyramids, and $[CuCl_3(H_2O)_3]$ groups. The presence of hydrogen selenites (Se—O—H) was confirmed by IR and Raman spectra. The particularity of $[Cu_{0.335}Se_{0.582}(HSeO_3)_2CuCl_3(H_2O)_3]$ is that it undergoes a phase transition on heating at 383 K. High temperature structure investigation of the new compound is in our future plans in order to confirm the nature of this transformation.

Acknowledgements

The author gratefully acknowledges the support of the University of Sfax.

Author details

Mohamed Loukil

Address all correspondence to: m.loukil@yahoo.fr

Sfax Faculty of Sciences - Sfax University, Laboratory Material Sciences and Environment, Tunisia

References

[1] Koskenlinna M, Valkonen J. Acta Crystallographica Section C. 1995;**51**:1637

[2] Micka Z, Cermak M, Niznansky D. Chemical Communications. 1990;**55**:2441

[3] Unterderweide K, Engelen B, Boldt K. Journal of Molecular Structure. 1994;**322**:233

[4] Boldt K. Thesis, Universita Siegen; 1994

[5] Valkonen J. Journal of Solid State Chemistry. 1986;**65**:363

[6] Hiltunen L, Leskela M, Niinisto L, Tammenmaa M. Acta Chemica Scandinavica. Series A. 1985;**39**:809

[7] Trombe JC, Lafront AM, Bonvoisin J. Inorganica Chimica Acta. 1997;**23**:847

[8] Lafront AM, Trombe JC, Bonvoisin J. Inorganica Chimica Acta. 1995;**238**:15

[9] Lafront AM. Thesis, University of Toulouse. 1995

[10] Sheldrick GM. SHELXS97. Program for the Refinement of Crystal Structures. Germany: University of Gottingen; 1986

[11] Sheldrick GM. SHELXL97. Program for the Refinement of Crystal Structures. Germany: University of Gottingen; 1997

[12] Shannon RD. Acta Crystallographica. Section A. 1976;**32**:751-767

[13] Cody CA, Levitt RC, Viswanath RS, Miller PJ. Journal of Solid State Chemistry. 1978;**26**:281

[14] Richtera L, Taraba J, Toužín J. Zeitschrift für Anorganische und Allgemeine Chemie. 2003; **629**:716-721

[15] Ratheesh R, Suresh G, Bushiri MJ, Nayar VU. Spectrochimica Acta A. 1995;**51**:1509

[16] Lutz HD, Haeuseler H. Journal of Molecular Structure. 1999;**511**:69-75

[17] Tanimizu M, Takahashi Y, Nomura M. Geochemical Journal. 2007;**41**:291

[18] Jahn HA, Teller E. Proceedings of the Royal Society of London, Series A. 1937;**161**:220

[19] Claydon MF, Sheppard N. Chemical Communications. 1969:1431

[20] Bartos S, Ratajczak H. Journal of Molecular Structure. 1982;**76**:77

[21] Baran J, Lis T, Marchewka M, Ratajczak H. Journal of Molecular Structure. 1991;**250**:13

[22] Ratajczak H, Yaremko AM, Baran J. Journal of Molecular Structure. 1992;**275**:235

[23] Kloprogge JT, Frost RL. Applied Spectroscopy Issue. 2000;**4**(54):517

Chapter 4

Thiophene *S*-Oxides

Thies Thiemann

Additional information is available at the end of the chapter

http://dx.doi.org/10.5772/intechopen.79080

Abstract

Thiophene *S*-oxides constitute a class of molecules that have been studied in more detail only recently. Their existence as intermediates in the peracid mediated oxidation of thiophenes to thiophene *S,S*-dioxides, however, has been known over some time. Over the last 20 years, a larger number of thiophene *S*-oxides have been prepared and isolated in pure form. Thiophene *S*-oxides have been found to be good dienes in [4 + 2]-cycloaddition reactions, where they react with electron-poor, electron-neutral and electron-rich dienophiles with high *syn* π-facial stereoselectivity. Thiophene *S*-oxides have been found to be metabolites of thienyl-containing pharmaceuticals such as the anti-platelet drugs ticlopidine and clopidogrel. The chapter gives an overview of the preparation and reactivity of this class of compounds.

Keywords: thiophenes, selective oxidation, cycloaddition, functionalized arenes, drug metabolites

1. Early history of oxidation reactions of thiophenes: cycloaddition reactions of thiophene *S*-oxides prepared *in situ* in absence of Lewis acids

In the first half of the 20th century, considerable effort was devoted to the oxidation of the heteroaromatic thiophene (**1**) with the understanding that the oxidation of thiophene to thiophene *S,S*-dioxide (**2**) (**Figure 1**) would be accompanied by the loss of aromaticity [1, 2]. The non-substituted thiophene *S,S*-dioxide (**1**) is not very stable in the pure state [3], but undergoes a slow dimerization with concurrent extrusion of SO_2 from the primary cycloadduct (**4**) [4], leading to **5** (**Scheme 1**). Only much later were the properties and reactivity of pure, isolated non-substituted thiophene *S,S*-dioxide (**2**) described [5].

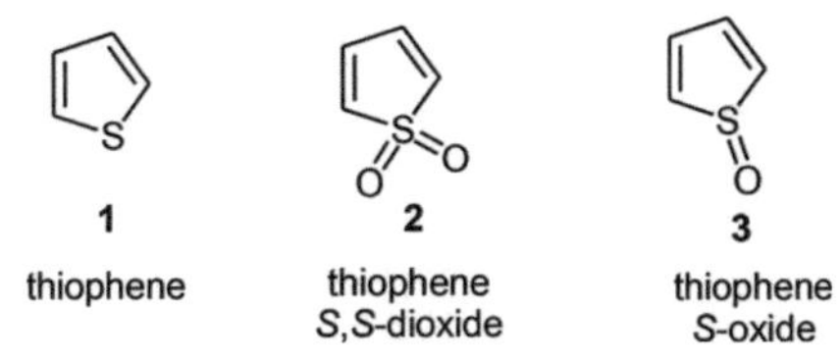

Figure 1. Structure of thiophene (**1**) and oxygenated thiophenes **2** and **3**.

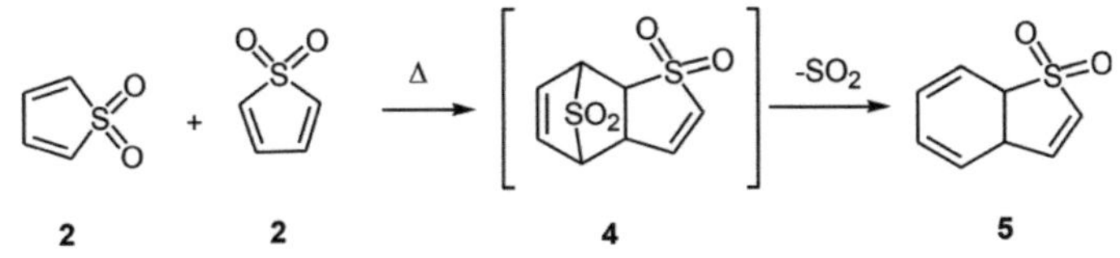

Scheme 1. Dimerisation of unsubstituted thiophene *S*,*S*-dioxide (**2**).

Much of the early work on the oxidation of thiophenes to thiophene *S*,*S*-dioxides involved hydrogen peroxide (H_2O_2) as oxidant, later *meta*-chloroperoxybenzoic acid (*m*-CPBA). That thiophene *S*-oxide was an intermediate in such oxidation reactions [6–8] was evident from the isolation of so-called sesquioxides as dimerization products of thiophene *S*-oxides [9–12]. Here, the thiophene *S*-oxide acted as diene with either another molecule of thiophene *S*-oxide or thiophene *S*,*S*-dioxide acting as ene [9–12] to give cycloadducts **6–8** (**Figure 2**). Thiophene *S*-monoxide (**3**) as an intermediate in the oxidation process of thiophene (**1**) to thiophene *S*,*S*-dioxide (**2**) could not be isolated under the conditions.

Nevertheless, the idea that a thiophene *S*-oxide intermediate could be reacted with an alkene of choice led Torssell [13] oxidize methylated thiophenes with *m*-CPBA in the presence of quinones such as *p*-benzoquinone (**12**). This gave cycloadducts **13** and **14** (**Scheme 2**) [13]. Further groups [11, 12, 14–19] used this strategy to react thiophene *S*-oxides such as **11**, prepared *in-situ* with alkenes and alkynes in [4 + 2]-cycloadditions (**Schemes 3** and **4**). In the reaction with alkenes, 7-thiabicyclo[2.2.1]heptene *S*-oxides such as **13** were obtained, while the reaction of thiophene *S*-oxides with alkynes led to cyclohexadienes and/or to aromatic products, where the initially formed, instable 7-thiabicyclo[2.2.1]hepta-2,5-diene *S*-oxide system **21** extrudes its SO bridge spontaneously (**Scheme 4**). A number of synthetic routes to multifunctionalized cyclophanes **32** [17], aryl amino acids **25** [16] and to crown ethers **29** [15] (**Scheme 5**)

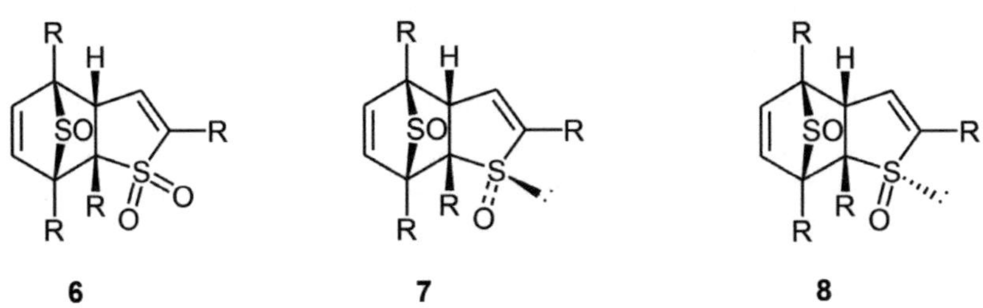

Figure 2. Sesquioxides obtained by dimerization of elusive thiophene *S*-oxide and by cycloaddition of thiophene *S*-oxide to thiophene *S*,*S*-dioxide.

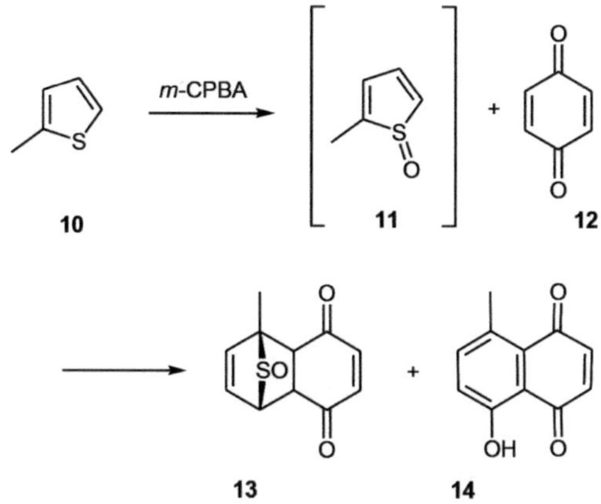

Scheme 2. Thiophene *S*-oxide (**11**), created *in situ*, reacts in Diels-Alder type fashion with *p*-benzoquinone (**12**).

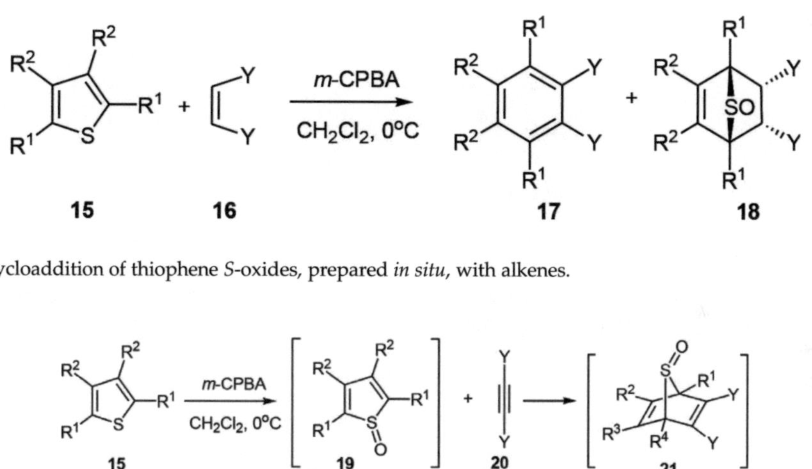

Scheme 3. Cycloaddition of thiophene *S*-oxides, prepared *in situ*, with alkenes.

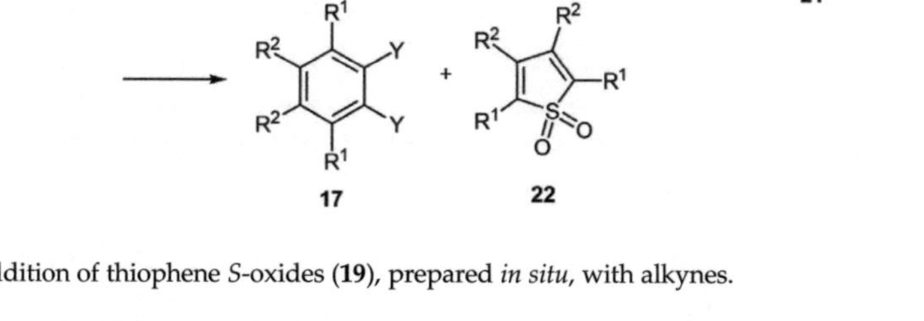

Scheme 4. Cycloaddition of thiophene *S*-oxides (**19**), prepared *in situ*, with alkynes.

have used the cycloaddition of thiophene *S*-oxides **19**, created *in-situ*, as a key step. The formation of the 7-thiabicyclo[2.2.1]heptene *S*-oxides (such as **13**, **18**) proceeds with stereocontrol. The cycloadditions yield predominantly *endo*-cycloadducts, with the oxygen of the sulfoxy bridge directed towards the incoming dienophile, exhibiting the *syn*-π-facial stereoselective nature of the reaction (see below for further discussion of the stereochemistry of the cycloadducts). Thiophene *S,S*-dioxides **2** possess an electron-withdrawing sulfone group, which leads both to a polarization and to a reduction of the electron density in the diene [20]. This results in a decrease of the energy of the HOMO as compared to identically

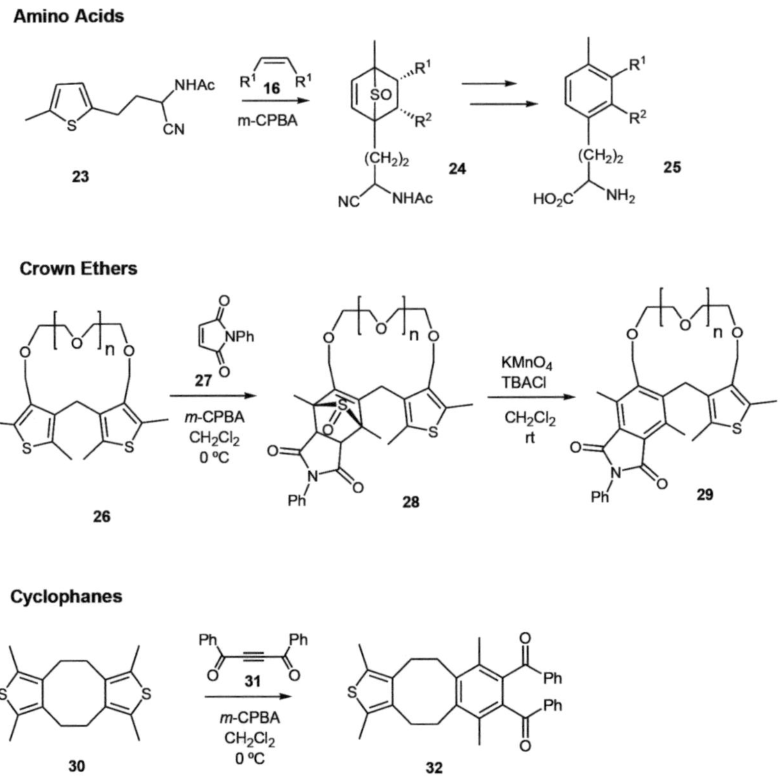

Scheme 5. Cycloaddition of thiophene *S*-oxides prepared *in situ*—applications in the synthesis of functionalized aminocarboxylic acids **25**, crown ethers **29** and cyclophanes **32**.

substituted cyclopentadienes [20]. Thiophene *S,S*-dioxides **2** are sterically more exacting than C_5 non-substituted cyclopentadienes, with the lone electron pairs on the sulfone oxygens leading to adverse non-bonding interactions with potentially in-coming dienophiles of high π-electron density. Thus, thiophene *S,S*-dioxides **2** often require higher temperatures [21, 22] in cycloaddition reactions than identically substituted cyclopentadienes. Recent frontier molecular orbital calculations at the HF/6-311++G(d,p)//M06-2X/6-31+G(d) level theory have shown that both HOMO (by 0.5 eV) and LUMO (by 0.4 eV) in thiophene *S*-oxide (**3**) are slightly higher in energy than in thiophene *S,S*-dioxide (**2**) [23].

Oxidation of the thienyl-unit in **33** leads to an intramolecular cycloaddition, where indanones **34** are obtained (**Scheme 6**) [24].

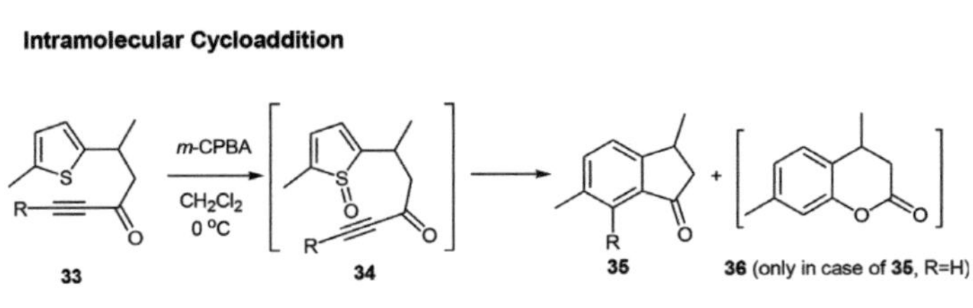

Scheme 6. Intramolecular cycloaddition of *in situ* prepared thiophene *S*-oxide **34**.

http://dx.doi.org/10.5772/intechopen.79080

2. Cycloaddition reactions of thiophene S-oxide prepared in situ in the presence of Lewis acids: thiophene S-oxides are isolated

Yields of cycloadducts have been found to be much higher, when oxidative cycloaddition reactions of thiophenes are carried out with *meta*-chloroperoxybenzoic acid (*m*-CPBA) or with H_2O_2 at lower temperatures such as at −20°C in the presence of a Lewis acid catalyst such as $BF_3{\cdot}Et_2O$ [11, 12, 25, 26] (**Scheme 7**) or of trifluoroacetic acid (CF_3CO_2H) [27]. Electron-poor dienophiles such as tetracyanoethylene, acetylene dicarboxylates, quinones, maleimides and maleic anhydride and mono-activated enes such as cyclopentenone and acrolein were used in these reactions.

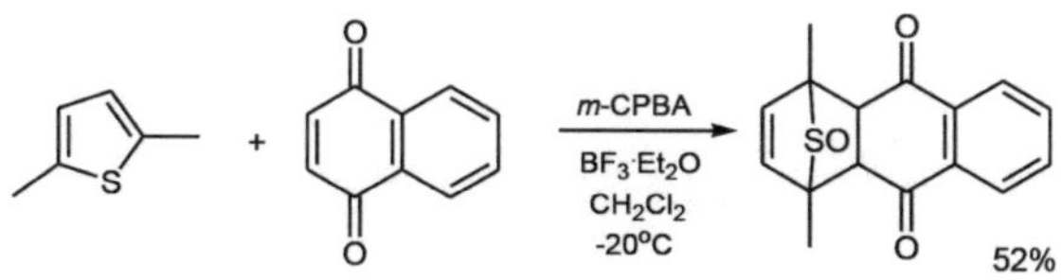

Scheme 7. Oxidative cycloaddition of thiophene **36** to naphthoquinone (**37**) in the presence of $BF_3{\cdot}Et_2O$.

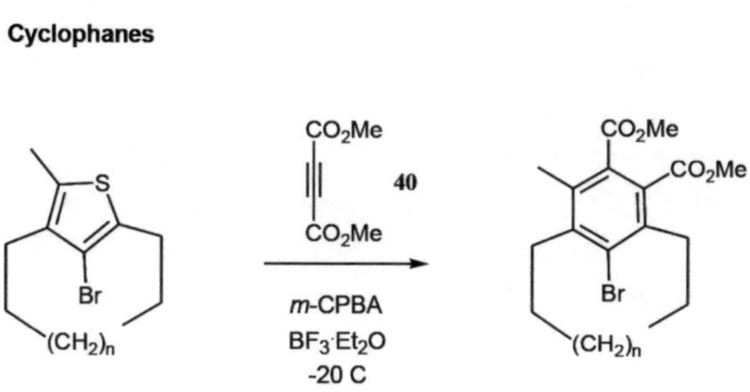

Scheme 8. Preparation of multifunctionalized cyclophane **41** by oxidative cycloaddition of thiophenophane **39** in the presence of $BF_3{\cdot}Et_2O$.

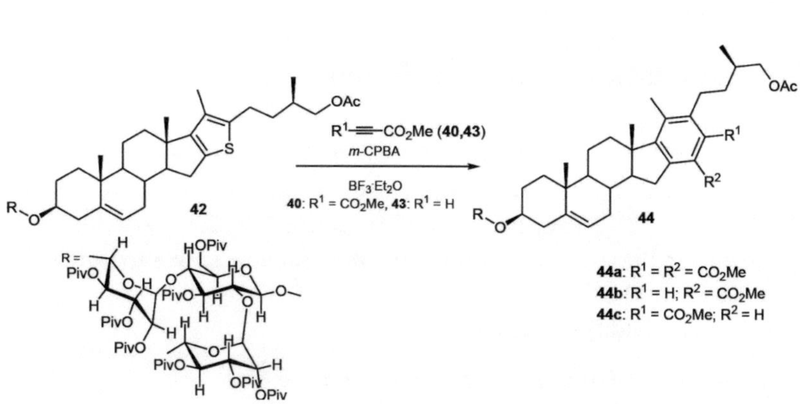

Scheme 9. Preparation of aethiosides A–C (**44a–c**) by oxidative cycloaddition of thienosteroidal sapogenin **42**.

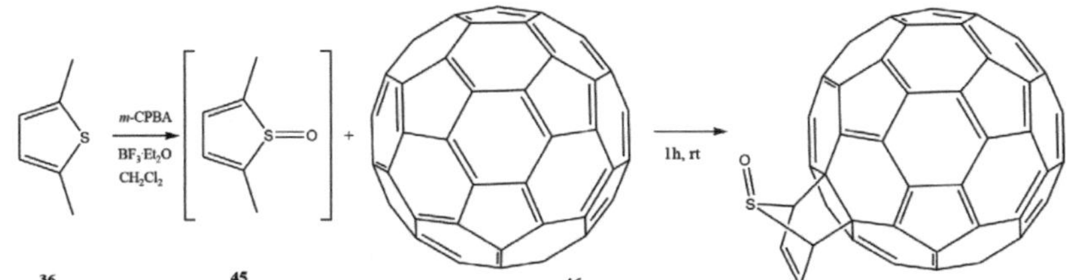

Scheme 10. Cycloaddition of 2,5-dimethylthiophene *S*-oxide (**45**), prepared *in situ*, to C60 (**46**).

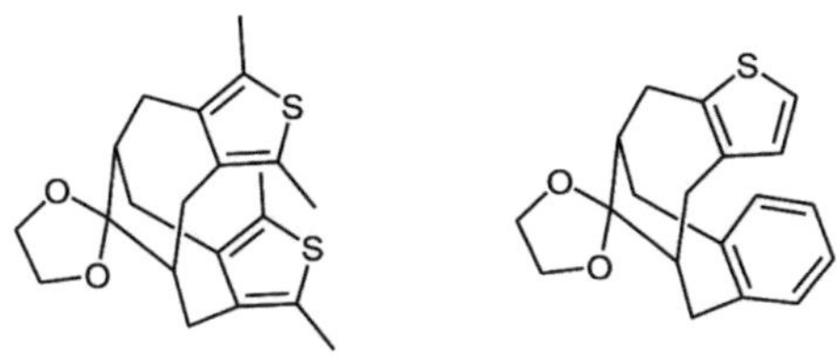

Figure 3. Orthothiophenophanes **48** and **49** do not allow for enough reaction volume and do not undergo oxidative cycloadditions with either alkynes or alkenes under the conditions (*m*-CPBA, $BF_3{\cdot}Et_2O$, CH_2Cl_2) [31].

Under the conditions *m*-CPBA/$BF_3{\cdot}Et_2O$, the cycloadditive transformation of thiophene *S*-oxides, prepared *in situ*, was used in the synthesis of new cyclophanes such as **39** (**Scheme 8**) [25]. A series of 2,3-bis(hydroxyphenyl) substituted 7-thiabicyclo[2.2.1]hept-2-ene S-oxides as potential estrogen receptor ligands were prepared by oxidative cycloaddition of 3,4-bis (hydroxyphenyl)thiophenes in the presence of $BF_3{\cdot}Et_2O$ [28]. Also the key step in Yu et al.'s [27] synthesis of steroidal saponins **44**, closely related to the E-ring areno containing natural products aethiosides A–C, is a $BF_3{\cdot}Et_2O$ catalyzed oxidative cycloaddition of the thieno-containing steroidal saponin **42** (**Scheme 9**) [26]. Furthermore, Zeng and Eguchi [29] were able to functionalize C60 (**46**) by cycloaddition with *in-situ* produced 2,5-dimethylthiophene S-oxide (**45**) [29, 30] (**Scheme 10**). Nevertheless, sterically hindered thiophenes are more difficult to be subjected to the oxidative cycloaddition reactions (**Figure 3**).

3. Preparation and isolation of pure thiophene *S*-oxides

Thiophene *S*-oxides could be isolated in pure form as side-products in a number of oxidative cycloaddition reactions using alkylated thiophenes as substrates run with *m*-CPBA in the presence of $BF_3{\cdot}Et_2O$ [11, 12]. Nevertheless, the first ascertained thiophene *S*-oxide (**51**) isolated in pure form came from the oxidation of the sterically exacting 2,5-bis-*tert*-butylthiophene (**50**) in absence of a Lewis acid or an added protic acid. 2,5-Bis-*tert*-butylthiophene *S*-oxide (**51**) could be isolated in 5% yield [32] (**Scheme 11**).

Previous to the isolation of thiophene *S*-oxides in pure form, based on UV-spectroscopic measurements, Procházka [33] had claimed that the parent thiophene *S*-oxide (**3**) could be

http://dx.doi.org/10.5772/intechopen.79080

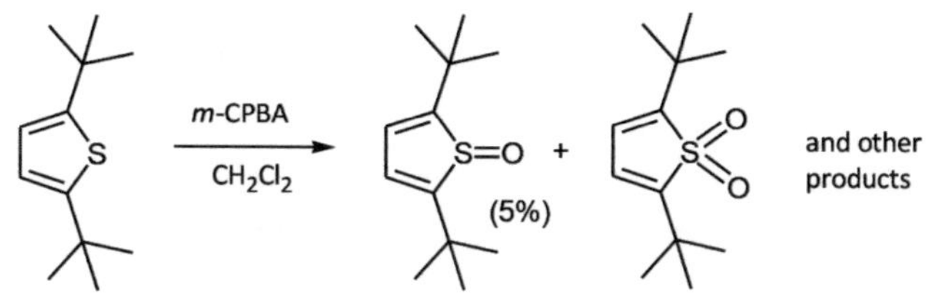

50 51 52

Scheme 11. Isolation of 2,5-bis-*tert*-butylthiophene *S*-oxide **51** by simple thiophene oxidation with *meta*-chloroperoxybenzoic acid (*m*-CPBA) [32].

prepared by double elimination from 3,4-dimesyloxy-2,3,4,5-tetrahydrothiophene *S*-oxide (**53**) and studied in solution. While subsequently the latter part of the assertion was thrown into doubt, the isolation of sesquioxides **7/8** from the reaction indicated at least the presence of thiophene *S*-oxide under these conditions [33] (**Scheme 12**).

Interestingly, a toluene solution of η5-ethyltetramethylcyclopentadienyl-η^4-tetramethylthienyl rhodium complex [Cp*Rh(η^4-TMT)] (**54**) can be oxidized with dry oxygen to [Cp*Rh(TMTO)] (**56**), which features a η^4-coordinated thiophene *S*-oxide ligand. Complex **56** was isolated and an X-ray crystal structure was carried out. Alternatively, [Cp*Rh(η^4-TMT)] (**54**) can be oxidized electrochemically to $[Cp^*Rh(\eta^4\text{-TMT})]^{2+}$ (**55**), which can also be obtained by protonation of [Cp*Rh(TMTO)] (**56**). Reaction of $[Cp^*Rh(\eta^4\text{-TMT})]^{2+}$ (**55**) with potassium methylsilanolate ($KOSiMe_3$) leads back to [Cp*Rh(TMTO)] (**56**) [34] (**Scheme 13**).

The reaction of the cationic transitory ruthenium complex $[Ru(C_6R_6)(C_4R_4S)]^+$ (**57**) with hydroxyl anion (OH^-) gives $Ru(C_6H_6)(C_4R_4SO)$ (**58**) [35] (**Scheme 14**). Here, in contrast to the complex [Cp*Rh(TMTO)] (**56**), the thiophene *S*-oxide ligand in $Ru(C_6H_6)(C_4R_4SO)$ (**58**) is not

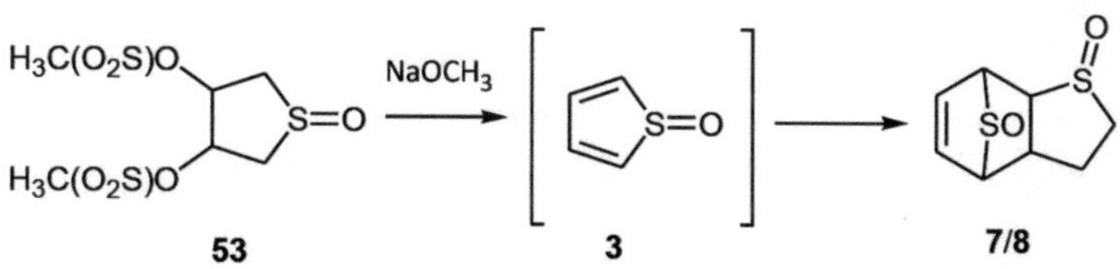

53 3 7/8

Scheme 12. *In situ* preparation of parent thiophene *S*-oxide (**3**) by an elimination reaction [33].

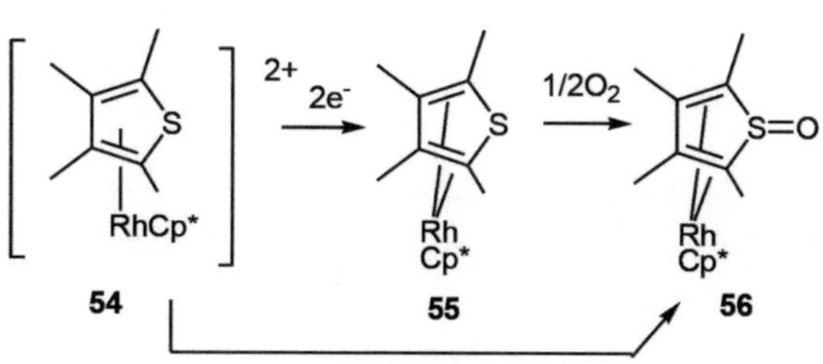

Scheme 13. Oxidation of [Cp*Rh(η^4-TMT)] (**54**) to [Cp*Rh(TMTO)] (**56**) [34].

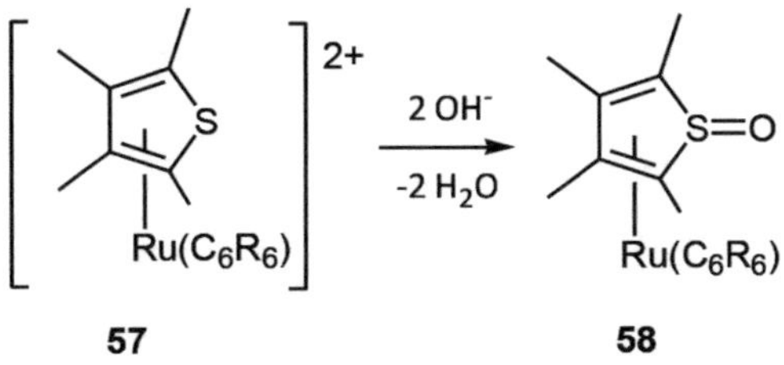

Scheme 14. Base hydrolysis of $[Ru(C_6R_6)(C_4R_4S)]^+$ (**57**) [34].

stable, but opens to an acetylpropenethiolate. Stable osmium thiophene *S*-oxide complexes of type (cymene)Os(C_4Me_4S=O) have also been prepared [36]. In neither of the cases, was it tried to decomplex the thiophene *S*-oxide ligand.

In the 1990s, two main synthetic methodologies were developed to prepare thiophene *S*-oxides **63**. The first involves the reaction of substituted zirconacyclopentadienes **62** with thionyl chloride ($SOCl_2$), developed by Fagan et al. [37, 38] and by Meier-Brocks and Weiss [39]. Typically, tetraarylzirconacyclopentadienes **62a** can be synthesized easily by reacting $CpZrCl_2$ (**59**), *n*-BuLi and diarylethyne (**61a**) in one step (**Scheme 15**). This strategy was followed by Tilley et al. [40, 41] in their synthesis of substituted thiophene *S*-oxides. Miller et al. published results for a synthesis of 2,5-diarylthiophene *S*-oxides (**63b**) along the same lines, using ethynylarene (**61b**) [42].

The other methodology involves an oxidation of a thiophene with either a peracid in the presence of a Lewis acid such as titanium tetrachloride ($TiCl_4$) [43] or boron trifluoride etherate ($BF_3{\cdot}Et_2O$) [44, 45] or with hydrogen peroxide in the presence of a protonic acid such as trifluoroacetic acid [46, 47] (**Scheme 16**). Also, the use of the reaction system H_2O_2 in presence of NaFe(III) ethylenediaminetetraacetate/Al_2O_3 has been reported [48, 49] (**Scheme 16**) as has been the use of the reaction system $[(C_{18}H_{37})_2(CH_3)_2N]_3[SiO_4H(WO_5)_3]$ [50]. The thiophene *S*-oxides **65**, suitably substituted, can be isolated by column chromatography and can be held in substance for a number of weeks without appreciable degradation, when in crystallized form and when kept in the dark. It is supposed that the Lewis acid not only activates the

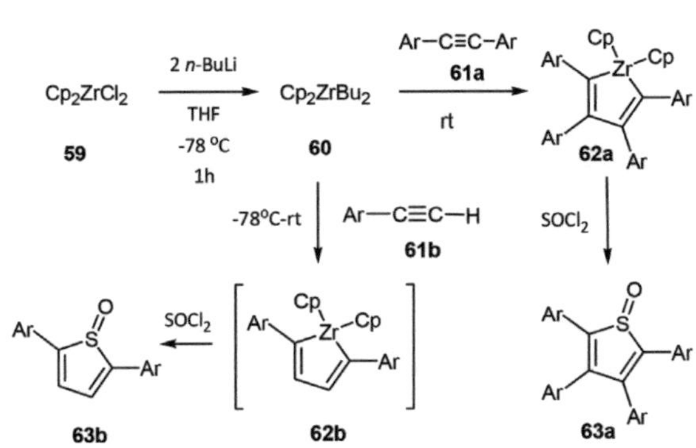

Scheme 15. Synthesis of tetraarylthiophene *S*-oxides **63a/b** by reaction of tetraarylzirconacyclopentadienes **62a/b** with $SOCl_2$.

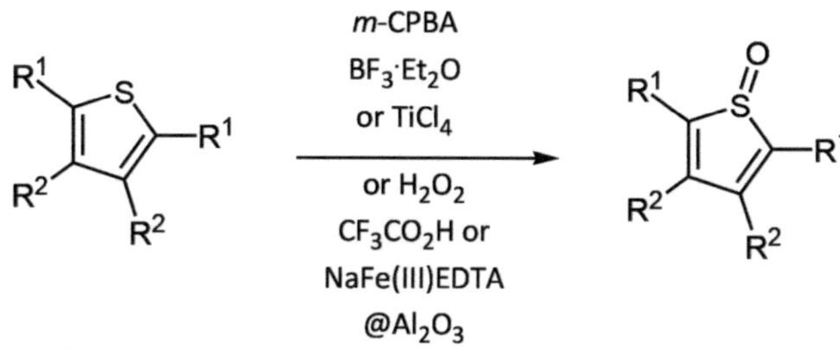

64 65

Scheme 16. Preparation of thiophene *S*-oxides **65** by oxidation of thiophenes **64** in the presence of a Lewis acid or a protonic acid.

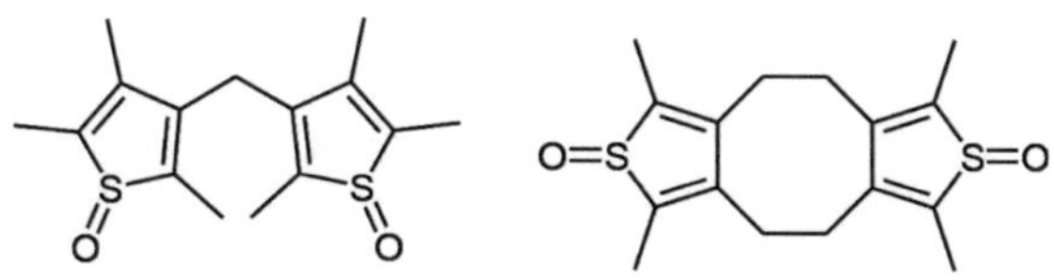

66 67

Figure 4. Known bisthienyl-*S*-oxides **66** and **67**.

peracid, but also coordinates to the oxygen in the formed thiophene *S*-oxide, thus reducing the electron-density on the sulfur of the thiophene *S*-oxide, making it less prone to undergo a second oxidation to the thiophene *S,S*-dioxide.

It has been shown that in a molecule, such as **66** or **67**, with two thienyl cores, both can be oxidized to thienyl-*S*-oxides with *m*-CPBA, $BF_3{\cdot}Et_2O$ CH_2Cl_2, −20°C) [11, 17] . Under these conditions, the second thiophene unit can compete successfully with a thiophene *S*-oxide for the oxidant (**Figure 4**).

4. Reactions of thiophene *S*-oxides

4.1. [4 + 2]-cycloaddition reactions

Even before thiophene *S*-oxides could be isolated in pure form, it was evident that thiophene S-oxides are good dienes in cycloaddition reactions, as "trapping" by cycloaddition reaction was one of the standard techniques to gauge the presence of thiophene *S*-oxide intermediates and provided a versatile preparative entry to 7-thiabi-cyclo[2.2.1]heptene *S*-oxides **68**. These in turn could be converted to substituted arenes **71** by either pyrolysis [15], photolysis [51], or PTC-catalyzed oxidative treatment with $KMnO_4$ [15] or electrochemical oxidation [18] or 7-thiabicyclo-[2.2.1]heptenes (**70**) by reaction of **68** with PBr_3 [52]. Reaction of **68** with tributyl-tin hydride gives cyclic dienes such as **72** [**—**X**—**X**—** = **—**(CO)N**—**Ph(CO)**—**]. Base catalyzed cleavage of the sulfoxy bridge of 1,4-dihalo-7-thiabicyclo[2.2.1]heptane *S*-oxides **68** (R^1 = Cl or Br) leads to the generation of diaryl disulfides such as **69** (**Scheme 17**).

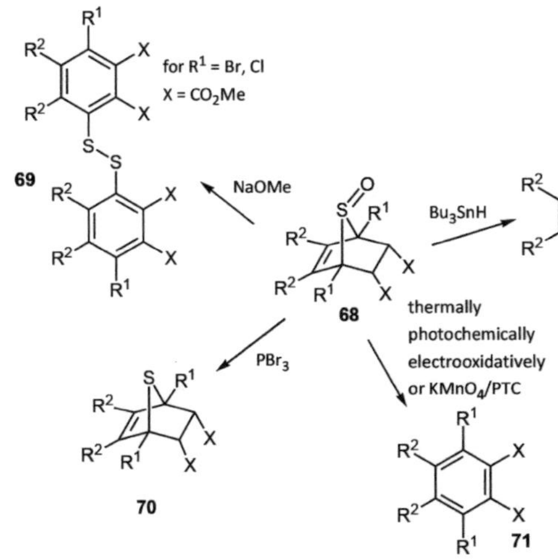

Scheme 17. 7-Thiabicyclo[2.2.1]heptene S-oxides **68** as versatile precursors to arenes.

With the possibility of isolating the thiophene *S*-oxides, it became possible to carry out cycloaddition reactions with alkenes that themselves react with *m*-CPBA. Thiophene *S*-oxides such as **73** have been found to react equally well with electron-rich alkenes such as enol ethers (**74**) [53], with electron neutral alkenes such as with cyclopentene (**76**) [53, 54] and with electron-poor alkenes such as with cyclopentenone or with maleic anhydride [11, 54] (**Scheme 18**). Also, thiophene *S*-oxides react with bicyclopropylidene (**82**) [55] under high pressure (10 kBar, **Scheme 19**), with allenes [56] (such as **79**, **Scheme 19**), with cyclopropylideneketone [55] (**Scheme 20**) and with benzyne (**90**) [56], both formed *in-situ* (**Scheme 21**). The reaction of tetrachlorocyclopropene (**93**) with 3,4-bis-*tert*-butylthiophene *S*-oxide (**73**) led to 6,7-bis-*tert*-butyl-2,3,4,4-tetrachloro-8-thiabicyclo[3.2.1]octa-2,6-diene 8-oxide (**95**), resulting from a ring opening of the primary cycloadduct **94** with a concomitant migration of a chloro atom [57] (**Scheme 22**). The ability of the thiophene *S*-oxides to undergo cycloadditions with alkenes, regardless of the electron demand of the reaction, has made Houk et al. say that *thiophene 1-oxide cycloadditions warrant their classification as click reactions* [23].

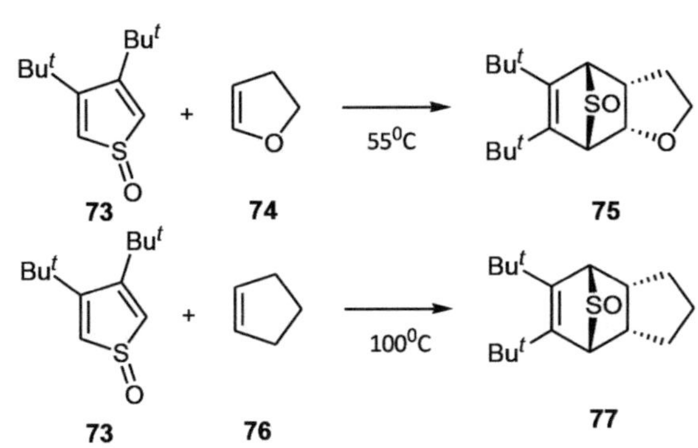

Scheme 18. 3,4-Bis-*tert*-butylthiophene *S*-oxide (**73**) cycloadding to electron-rich and electron-neutral alkenes.

http://dx.doi.org/10.5772/intechopen.79080

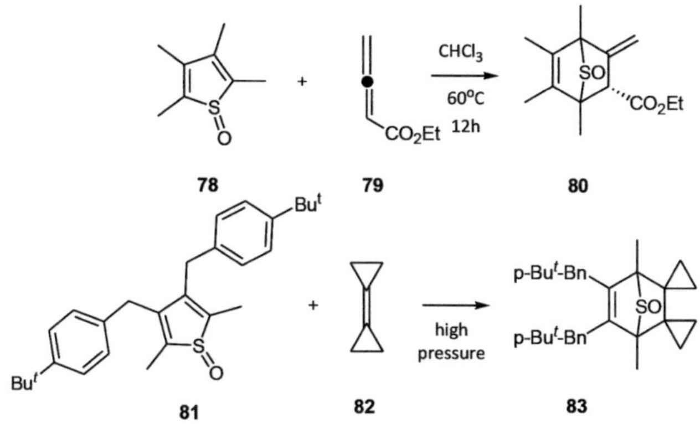

Scheme 19. Thiophene *S*-oxides cycloadd to allenes and to bicyclopropylidene (**82**) under high pressure.

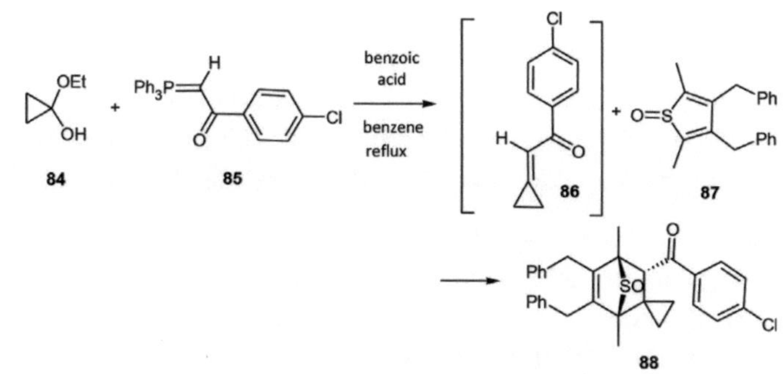

Scheme 20. One pot Wittig reaction—Diels Alder reaction with thiophene *S*-oxide **87** as diene.

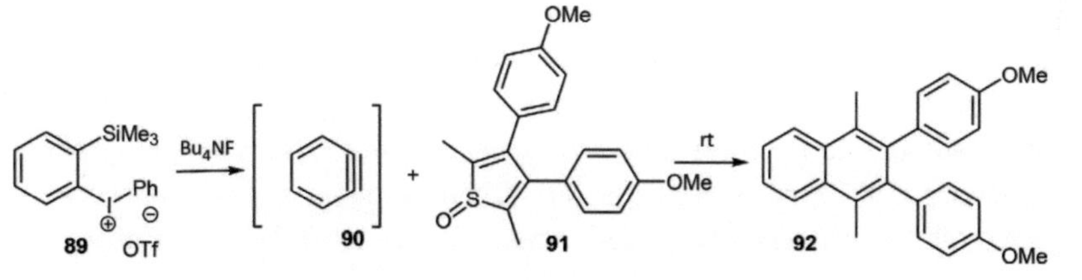

Scheme 21. Cycloaddition of thiophene *S*-oxide **91** with benzyne (**90**), prepared *in situ*.

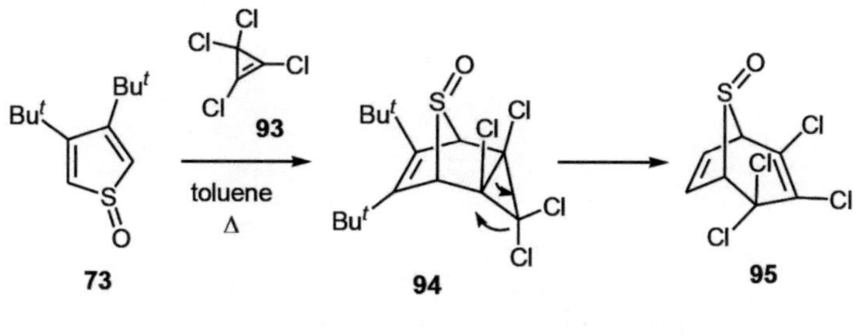

Scheme 22. Cycloaddition of thiophene *S*-oxide (**73**) with tetrachlorocyclopropene (**93**).

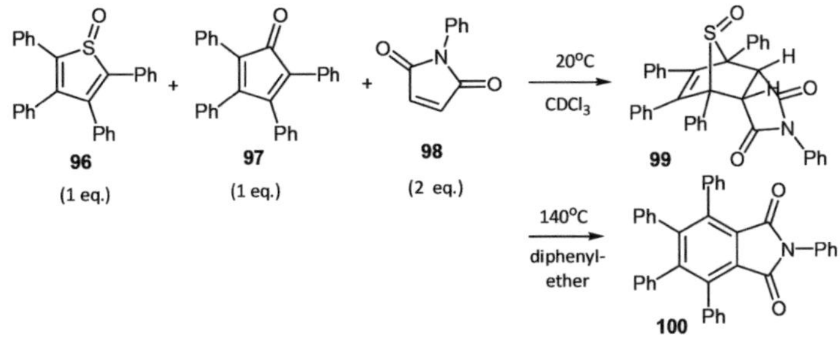

Scheme 23. Thiophene *S*-oxide **96** competes efficiently with tetracyclone **97** for *N*-phenylmaleimide (**98**).

Thiophene *S*-oxides are good precursors for the preparation of heavily substituted arenes such as **100** [58] (**Scheme 23**). Often, tetraarylcyclopentadienones **97** are used to synthesize oligoaryl benzenes by cycloaddition reaction. However, tetraphenylthiophene *S*-oxide (**96**) is the more reactive diene when compared to tetraphenylcyclopentadienone (**97**) as can be seen in the competitive cycloaddition of **96** and **97** with *N*-phenylmaleimide (**98**), where at room temperature only tetraphenylthiophene *S*-oxide undergoes cycloaddition to give **99** (**Scheme 23**) [58]. **99** can be converted to the heavily substituted phthalimide **100** [58], either by extruding the SO group thermally in diphenyl ether (**Scheme 23**) or by reaction with $KMnO_4$/PTC.

Sometimes, tetraphenylthiophene *S*-oxide (**96**) and tetraphenylcyclopentadienone (**97**) give different products in cycloaddition reactions. A typical example is their cycloaddition to benzo[*b*]thiophene *S,S*-dioxide (**101**), where the reaction with **96** leads to the formation of dibenzothiophene *S,S*-dioxide **102**, but with **97** gives dibenzothiophene **104** [59] (**Scheme 24**). The reason for this difference lies in the tendency of tetracyclines such as **94** to be oxidized to pyrones **102** at higher reaction temperatures, with the *S,S*-dioxides playing the oxidizing agent [59] (**Scheme 24**).

Again, cycloaddition reactions of purified thiophene *S*-oxides can be used to prepare multifunctionalized arenes such as cyclophanes (**Scheme 25**) [25]. Nakayama et al. [61] have used thiophene *S*-oxides to prepare sterically over freighted anthraquinones. Thiemann et al. [62] used halogenated thiophene *S*-oxides, albeit prepared *in-situ* to synthesize halogenated anthraquinones, which can easily be transformed further to arylated anthraquinones [63, 64].

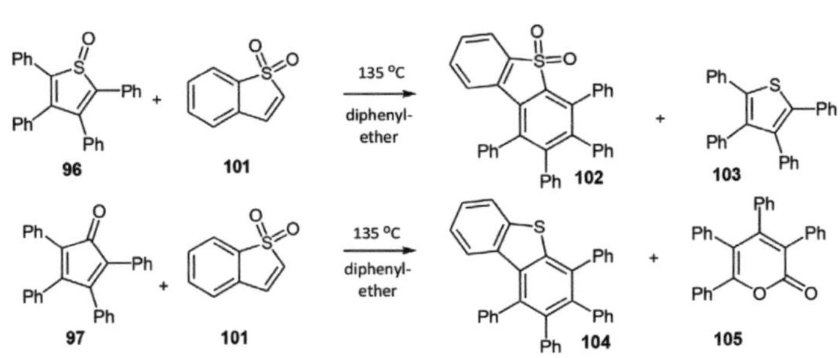

Scheme 24. Comparison of the cycloaddition of tetraphenylthiophene *S*-oxide **96** and tetracyclone **97** with benzo[*b*] thiophene *S,S*-dioxide (**101**). Tetracyclone **97** gives pyrone **105** as side product [59, 60].

http://dx.doi.org/10.5772/intechopen.79080

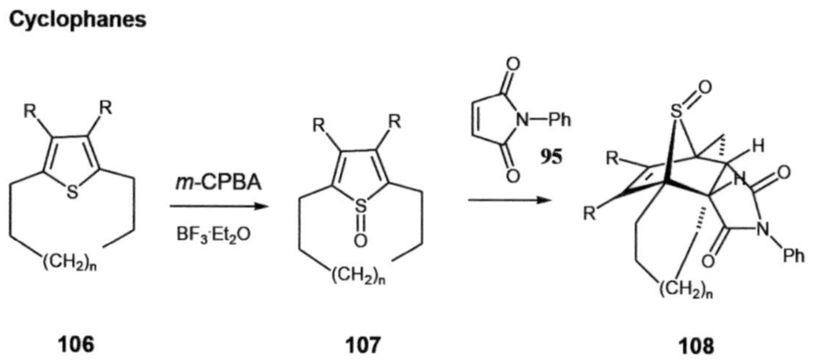

Scheme 25. Multifunctionalized cyclophanes **108** by cycloaddition of thiophenophane *S*-oxides **106**.

The cycloaddition reactions of purified thiophene *S*-oxides can be combined with other transformations in one pot, such as with Wittig olefination reactions (**Scheme 20**) [55].

Not all thiophene *S*-oxides undergo cycloaddition reactions with alkynes or alkenes. In general, appreciable reaction volume is needed to allow for the forming sulfoxy-bridge in the primary cycloadducts and, in some cases, of the subsequent extrusion of SO. Also, when considerable strain is associated with the thiophene *S*-oxides and/or the cycloadducts, reactions other than cycloadditions can occur. Thus, strained thiophenophane *S*-oxide **110** does not undergo a cycloaddition with **98**, but undergoes a rearrangement leading to oxygen insertion into the ring with concomitant extrusion of sulfur, leading to furanophane **111** (**Scheme 26**) [25]. Fujihara et al. were able to prepare the thiacalixarene *S*-oxide **112**; again, the thiacalixarene *S*-oxide did not undergo a cycloaddition reaction with alkyne **113**, but rather formed the thiophene-S,C-sulfonium ylide **114** (**Scheme 27**) [65].

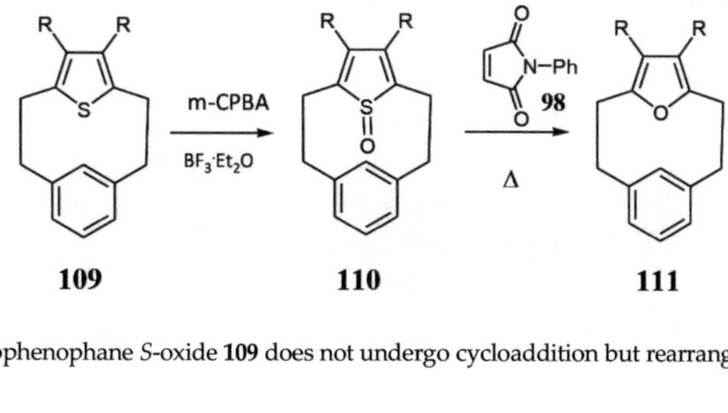

Scheme 26. [2.2]Metathiophenophane *S*-oxide **109** does not undergo cycloaddition but rearranges to [2.2]furanophane **111**.

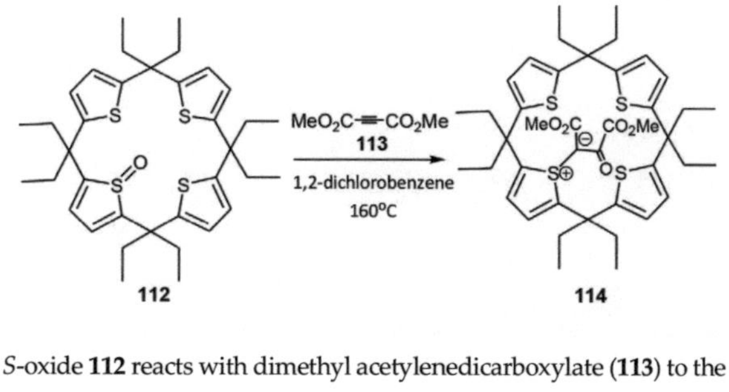

Scheme 27. Thiacalixarene *S*-oxide **112** reacts with dimethyl acetylenedicarboxylate (**113**) to the thiacalixarene *S*,*C*-ylide **114**.

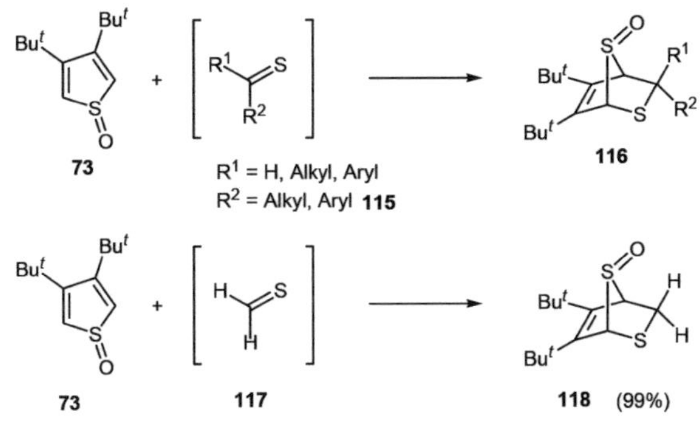

Scheme 28. Hetero-Diels-Alder reactions of 3,4-bis-*tert*-butylthiophene *S*-oxide (**73**).

Thiophene *S*-oxides as cyclic dienes undergo hetero-Diels-Alder reactions, also (**Scheme 28**). Thus, Nakayama et al. could establish that 3,4-bis-*tert*-butylthiophene *S*-oxide **73** reacts with thioaldehydes **115/117** and thioketones **115**, generated *in-situ* to give 2,7-dithiabicyclo[2.2.1] hept-5-ene 7-oxides **116** and **118** [66] (**Scheme 28**). The cycloadducts are *endo*-products as ascertained by X-ray crystallography and ^{1}H NMR spectroscopy. Thiobenzophenone could be reacted with good yield; however, here two isomeric products are produced, the major product originating from the *syn*-π-face while the lesser product from the *anti*-π-face cycloaddition.

Finally, **73** reacts with carbonyl cyanide [**121**, $CO(CN)_2$], created *in-situ* by oxidation of tetracyanoethylene oxide (**119**, TCNO) with thiophene *S*-oxide **73**, in *hetero*-Diels-Alder fashion to give **122** [67] (**Scheme 29**).

Nakayama et al. have calculated that the cycloadditions of the thiophene *S*-oxides are inverse electron demand reactions [53]. All of the above cycloaddition reactions are highly stereoselective, regardless whether the thiophene *S*-oxide is prepared and used *in-situ* or an isolated thiophene *S*-oxide is used. It is known that the thiophene *S*-oxides invert at the sulfur and inversion barriers have been calculated and measured experimentally for a number of these compounds [32, 68, 69]. Nevertheless, the sulfoxy group in the 7-thiabicyclo[2.2.1] heptene *S*-oxide systems is configurational stable. All the cycloadducts are *endo*-products.

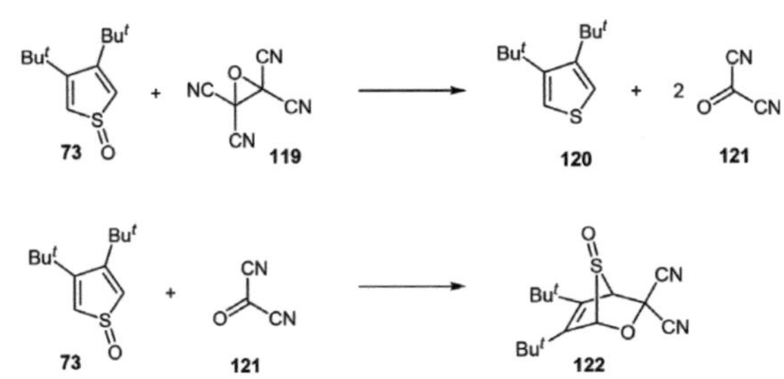

Scheme 29. Reaction of 3,4-tert-butylthiophene S-oxide (**73**) with tetracyanoethylene oxide (**119**, TCNO) and hetero-Diels Alder reaction to carbonyl cyanide (**121**).

In the cases where Lewis acids are used at low temperatures, this in itself is not surprising as it is known that low temperatures kinetically controlled cycloadducts are favored. Moreover, it has been stated that Lewis acid catalysis increases the extent of *endo*-addition in Diels-Alder reactions [70, 71]. The cycloadditions are seen to have *syn*-π-facial in that the dienophile adds *syn* to the oxygen. This means that the lone pair of the sulfur is directed towards the side of the newly formed double bond of the cycloadduct. A number of explanations have been given for the π-facial selectivity. Thus, Nakayama et al. rationalized that in the transition state less geometric change of the SO function would be required to reach the *syn*- rather than the *anti*-transition state geometry [53]. Also, a destabilizing interaction between the HOMO of the dienophile and the sulfur lone pair was noted in the *anti*-transition state [72]. The π-facial selectivity has also been explained by the Cieplak effect [73–75]. This effect was first proposed to account for the directing effect of remote substituents in addition reactions to substituted cyclohexanones. A large number of experimental observations in Diels-Alder reactions of dienophiles with 5-substituted cyclopentadienes have shown that the dienophiles will approach *anti* to the antiperiplanar σ-bond that is the better donor at the 5-position of the cyclopentadiene [76]. This σ-bond will best stabilize the σ-bonds formed in the transition state. Cycloadditions to thiophene *S*-monoxides have been predicted to occur *anti* to the lone electron-pair on sulfur, which is the better hyper-conjugative donor when compared to the oxygen of the sulfoxy-moiety. The lone pair electron orbital at the sulfur will stabilize the vacant σ*-orbitals of the developing incipient σ-bonds better than any orbital associated with the oxygen of the sulfoxy moiety [77] (**Figure 5**). This would be even more so, when the oxygen of the sulfoxy-unit is complexed by $BF_3{\cdot}Et_2O$.

Based on DFT computational studies, Houk et al. [23] showed that the ground state geometry of a thiophene *S*-oxide already resembles the molecule in its *syn* transition state. This distortion from planarity of the molecule minimizes its potential antiaromaticity which would result from a hyperconjugative effect by an overlap of $\sigma^*_{S=O}$ with the π-system (see also above/below) [23] (**Figure 6**).

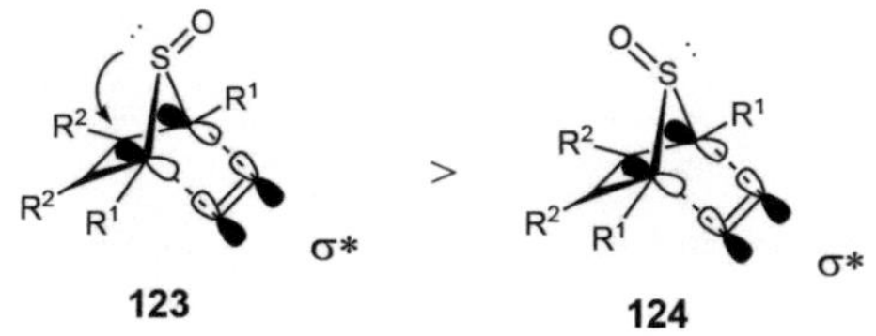

Figure 5. Transition state **123** preferred over transition state **124**.

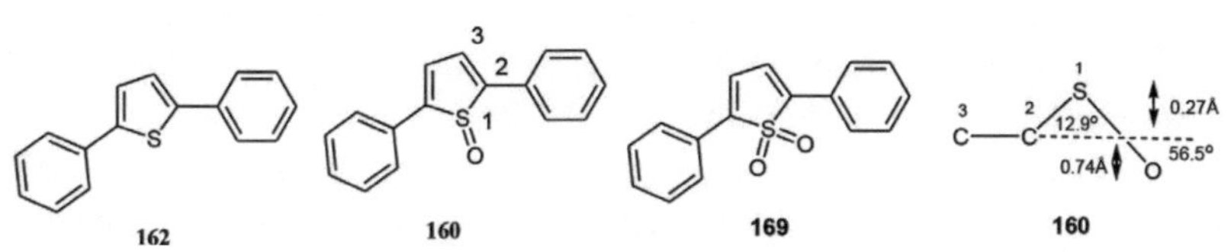

Figure 6. Structural feature of thiophene S-oxide **160**.

4.2. Further cycloaddition reactions

When heated with 2-methylene-1,3-dimethylimidazoline (**125**), 3,4-bis(*tert*-butyl)thiophene *S*-oxide **73** undergoes a [4π + 4π]-cycloaddition to the head-to-head dimer **126** (**Scheme 30**) [78]. Oxidation of the two sulfoxy bridges to sulfone **127** with dimethyldioxirane as oxidant is followed by thermally driven extrusions of the SO_2 bridges in **127** and gives 1,2,5,6-tetra(*tert*-butyl)octatetraene **128** [79] (**Scheme 30**).

Thiophene *S*-oxides react as enes in 1,3-dipolar cycloaddition reactions. Thus, 3,4-bis-*tert*-butylthiophene *S*-oxide (**73**) reacts with pyrroline *N*-oxide (**129**) to give cycloadduct **130** (**Scheme 31**) [80]. Nakayama et al. could show that **73** reacts with nitrile oxides, diazomethane, nitrile imides, nitrones, and azomethine ylides in *syn*-π-facial fashion [80].

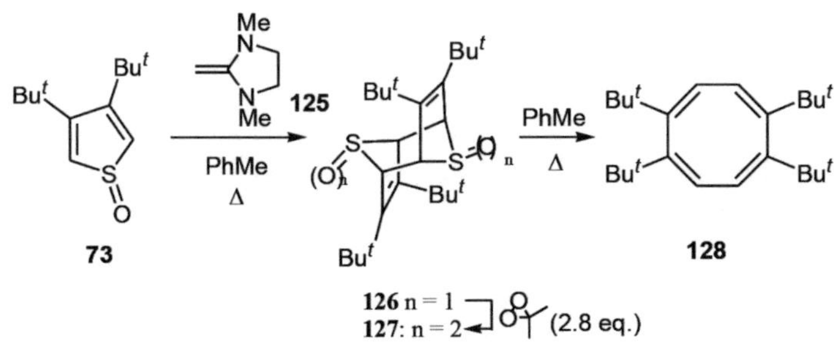

Scheme 30. [4π + 4π]-cycloaddition of thiophene *S*-oxide (**73**) to dimer **126**.

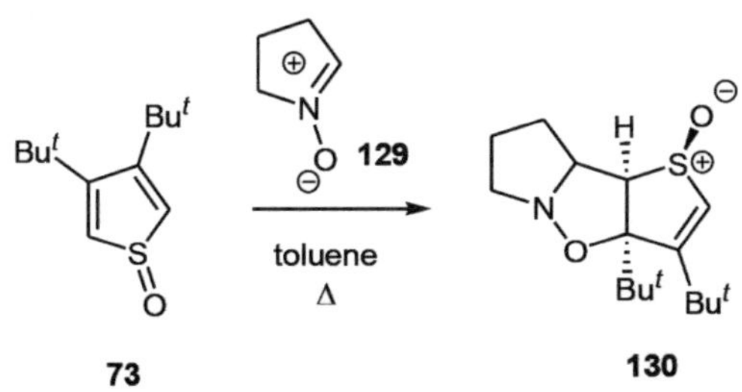

Scheme 31. [3 + 2]-cycloaddition of thiophene *S*-oxide (**73**) with pyrroline *N*-oxide (**129**) as 1,3-dipole.

4.3. Additions to thiophene *S*-oxides and other reactions

1,4-Additions are known for both 3,4-disubstituted and 2,5-disubstituted thiophene *S*-oxides [81–83]. Thus, bromine adds *cis* to both 3,4-bis-*tert*-butylthiophene *S*-oxide (**73**) [81] and 2,5-bis-trimethylsilylthiophene *S*-oxide (**134**) [82] to give the 2,5-dibromo-2,5-dihydrothiophene S-oxide derivatives **131** and **135** (**Scheme 32**). 3,4-Bis-*tert*-butylthiophene *S,S*-dioxide (**132**) undergoes *cis*-1,4-bromination, too [81] (**Scheme 32**). Also, alcohols and mercaptans have been submitted successfully to 1,4-additions with 3,4-bis-*tert*-butyl thiophene *S*-oxide (**73**) (**Scheme 33**) [83]. Interestingly, disulfur dichloride (S_2Cl_2) could be added to thiophene *S*-oxide **73**, leading to the rapid formation of adduct **137** (**Scheme 34**) [84]. **137**, however, is not stable

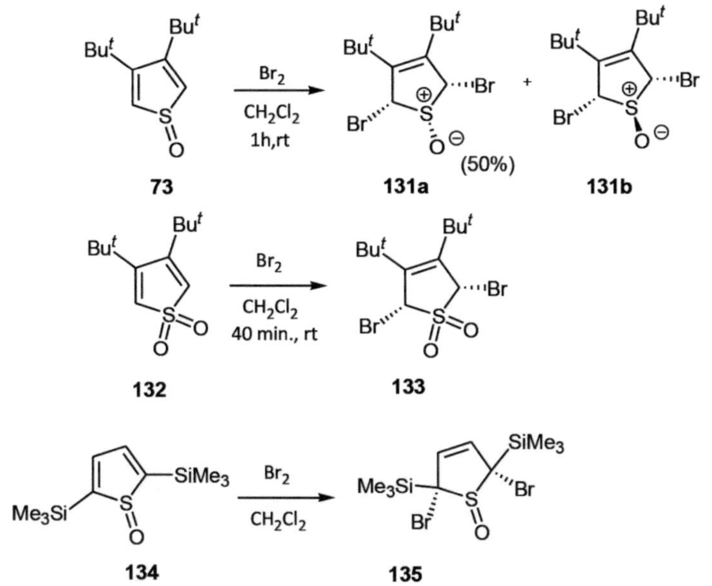

Scheme 32. Bromination of thiophene *S*-oxides **73** and **134** and thiophene *S,S*-dioxide **132**.

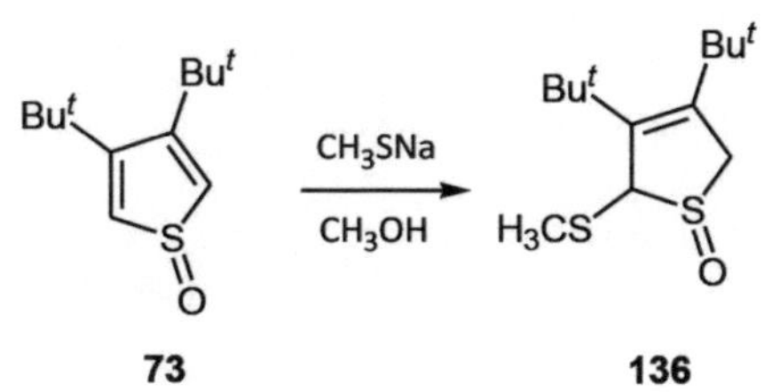

Scheme 33. Addition of methylthiolate to thiophene *S*-oxide (**73**).

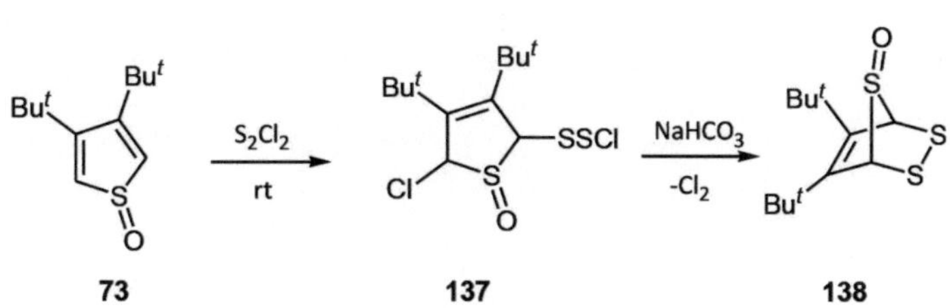

Scheme 34. Addition of disulfur dichloride (S_2Cl_2) to thiophene S-oxide **73**.

and transforms into **138**. **138** can be obtained with a 98% yield, when **137** is treated with aq. $NaHCO_3$ (**Scheme 34**) [84].

The sulfoxy group in thiophene *S*-oxide can be transformed into a sulfilimine or a sulfoximine moiety [85–87]. When thiophene *S*-oxide **73** is reacted with trifluoroacetic acid anhydride or triflic anhydride at −78°C, a mixture of sulfonium salt **139** and sulfurane **140** forms, which can be reacted with *p*-toluenesulfonamide (**141**) to provide, as the reaction mixture warms to room temperature, sulfilimine **142** (**Scheme 35**) [85, 86]. Sulfoximine **145** could be prepared by action of *N*-[(*p*-tolylsulfonyl)imino]phenyliodinane (TsN=IPh, **144**) on 2,4-bis-*tert*-butylthiophene *S*-oxide (**143**) in the presence of $Cu(CH_3CN)_4PF_6$ as catalyst. Further reaction of **145** with H_2SO_4 leads to *N*-unsubstituted sulfoximine **146** (**Scheme 36**) [86].

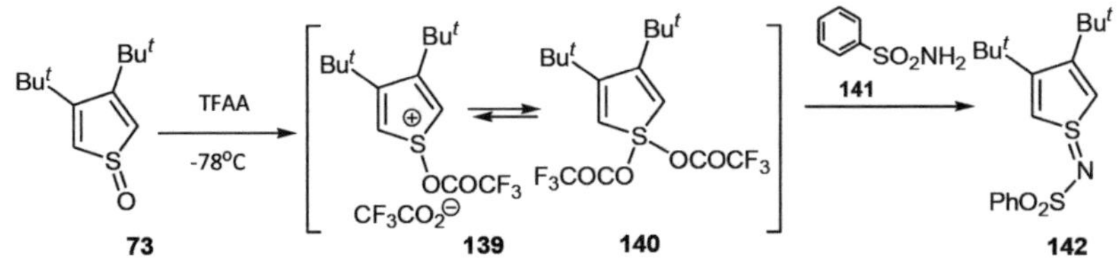

Scheme 35. Preparation of thiophene *S*-imide **142** from thiophene *S*-oxide **73**.

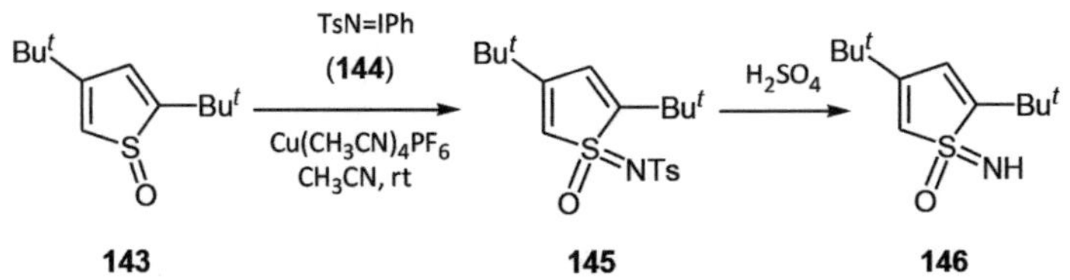

Scheme 36. Thiophene sulfoximines **145** and **146** from thiophene *S*-oxide **143**.

4.4. Photochemistry of thiophene *S*-oxides

The photochemical deoxygenation of dibenzothiophene *S*-oxides has been studied for quite some time [88–91] and has been found to proceed via the release of ground state atomic oxygen [$O(^3P)$] upon photoirradiation (**Scheme 37**). Thiophene *S*-oxides deoxygenate photochemically as well. Nevertheless, the photochemistry of thiophene *S*-oxides is intrinsically more complex than that of dibenzothiophene *S*-oxides, often providing a mixture of products, depending on the substitution pattern of the photoirradiated thiophene *S*-oxide. The photolysis of 2,5-bis(trimethylsilyl)thiophene *S*-oxide (**134**) leads exclusively to deoxygenation to

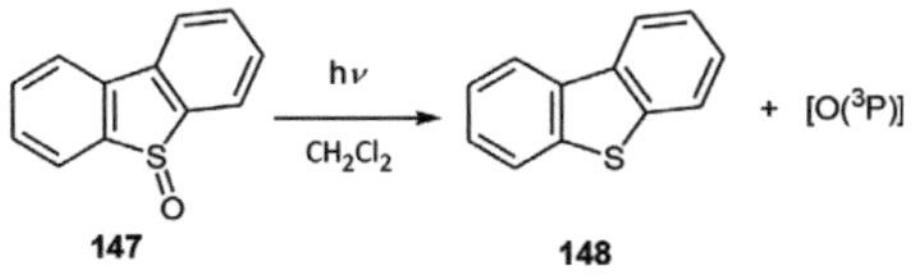

Scheme 37. Photodeoxygenation of dibenzothiophene *S*-oxide (**147**).

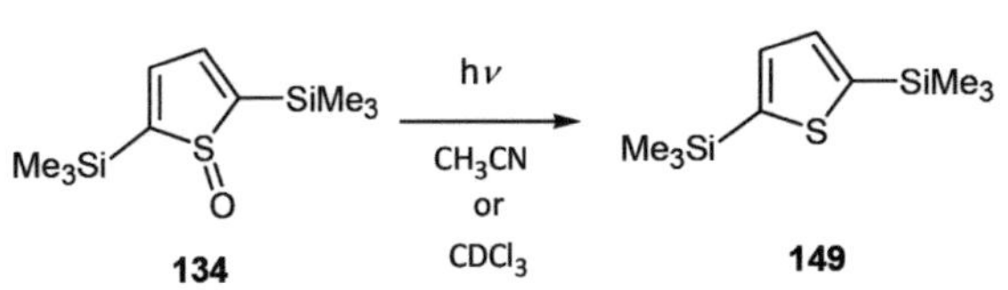

Scheme 38. Photolysis of 2,5-bis(trimethylsilyl)thiophene *S*-oxide (**134**).

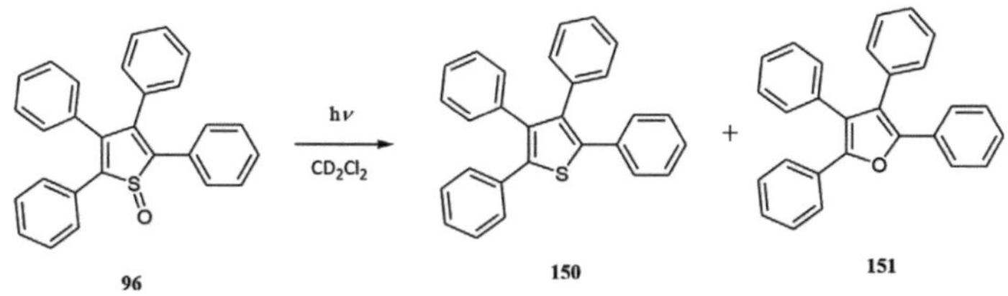

96 150 151

Scheme 39. Photolysis of tetraphenylthiophene *S*-oxide (**96**).

produce 2,5-trimethylsilylthiophene (**149**) (**Scheme 38**). Otherwise, in those cases, where the thiophene *S*-oxide does not exhibit a CH_3 substituent on the ring system, furans are often the main products along with (deoxygenated) thiophenes (**Scheme 39**). This has been noted with phenyl-substituted (**96**, **160**) and *tert*-butyl substituted thiophene *S*-oxides (**73**, **143**, **153**) as well as with 3,4-dibenzylthiophene *S*-oxide (**158**) (**Scheme 40**) [92–95]. Different mechanisms have been forwarded for this photochemical formation of furans. A viable mechanism involves a cyclic oxathiin, where the first step within the photochemical reaction is initiated by the homolytic ring cleavage α to the sulfoxy group [92–94]. A rearrangement of thiophene *S*-oxides to produce furans can also proceed thermally as found by Thiemann et al. [18] in the transformation of thiophenophane *S*-oxide **110** to furanophane **111** (**Scheme 26**) and by

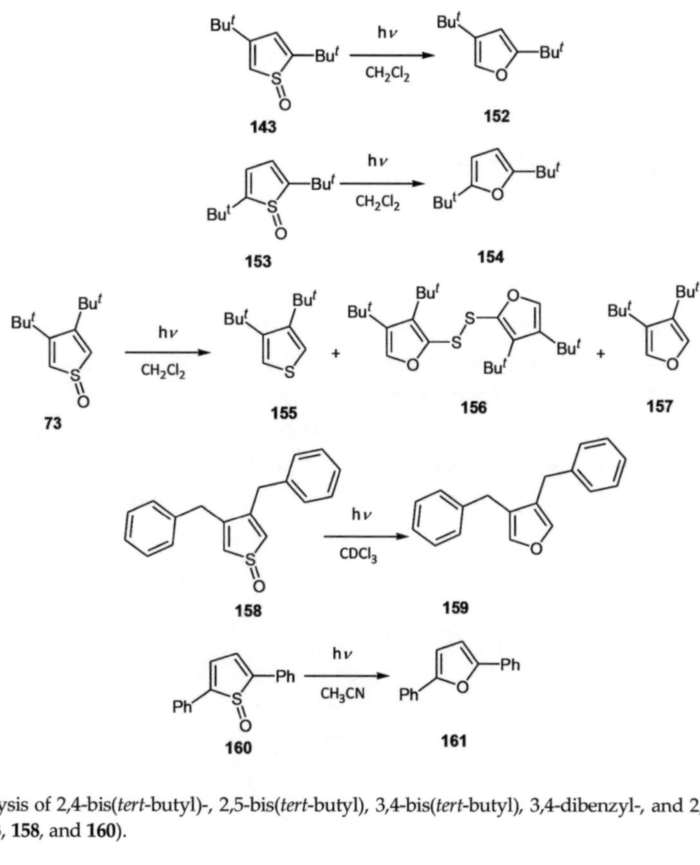

143 152 153 154 73 155 156 157 158 159 160 161

Scheme 40. Photolysis of 2,4-bis(*tert*-butyl)-, 2,5-bis(*tert*-butyl), 3,4-bis(*tert*-butyl), 3,4-dibenzyl-, and 2,5-diphenylthiophene *S*-oxide (**143**, **153**, **73**, **158**, and **160**).

Mansuy, Dansette et al. in their oxidation of 2,5-diphenylthiophene (**162**) with H_2O_2/CF_3CO_2H to 2,5-diphenylthiophene *S*-oxide (**163**), where an appreciable amount of furan **164** was formed as side-product [46] (**Scheme 41**). In the case of methyl substituted thiophene *S*-oxides, hydroxyl-alkylthiophenes such as **166** and follow-up products such as ether **167** have been isolated as photoproducts [96] (**Scheme 42**).

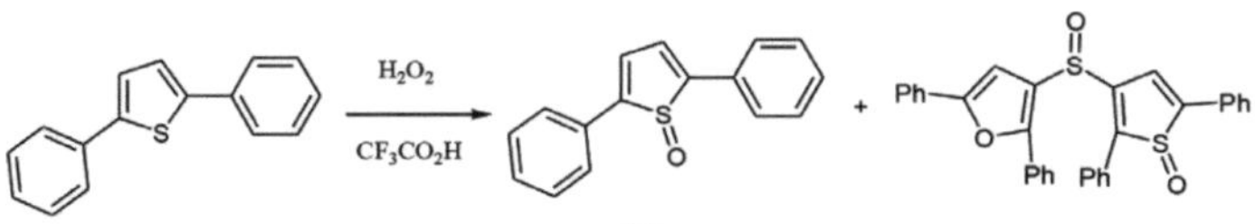

Scheme 41. Formation of furan **163** in the oxidation of 2,5-diphenylthiophene (**162**).

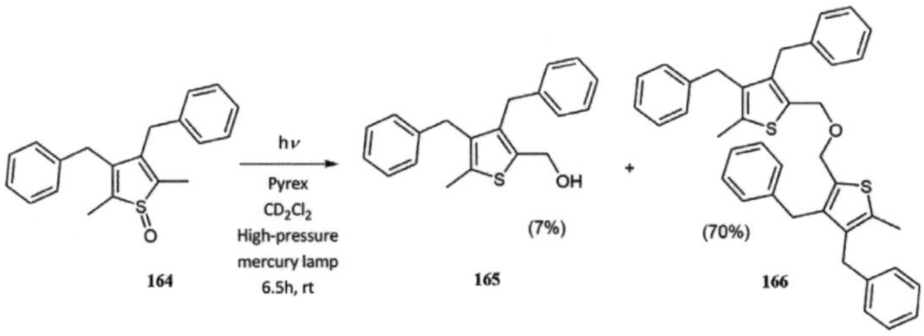

Scheme 42. Photolysis of 3,4-dibenzyl-2,5-dimethylthiophene *S*-oxide (**165**).

4.5. Electrochemistry of thiophene *S*-oxides

Thiophene *S*-oxides such as **164** and **167** show well-defined, chemically irreversible CV reduction waves, where two reduction processes seem to compete. In the presence of a proton donor, the reduction waves experience a significant shift to more positive potentials, although the reduction potential is still dependent on the substitution pattern of the thiophene *S*-oxides [96]. In the presence of a proton donor such as benzoic acid at higher concentrations, the reduction of a thiophene *S*-oxide such as of **167** becomes a straightforward two proton—two electron reduction process to the corresponding thiophene [96]. Bulk electrolysis of thiophene *S*-oxides in presence of 10-fold excess of benzoic acid has been carried out and have led to the corresponding thiophenes in up to 90% isolated yield (**Scheme 43**) [96]. Also, thiophene

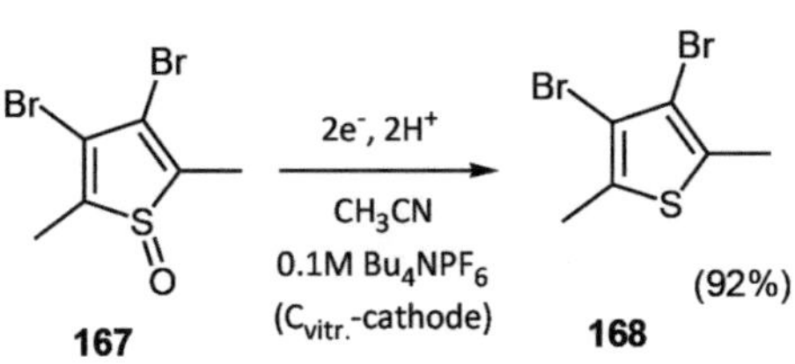

Scheme 43. Electrochemical reduction of 3,4-dibromo-2,5-dimethylthiophene *S*-oxide (**167**) in the presence of 10 eq. benzoic acid.

S-oxides show oxidative electrochemistry at platinum in MeCN/Bu_4NPF_6 [97]. The electrochemical oxidation of tetraphenylthiophene *S*-oxide under the above conditions leads mainly to the formation of diphenylacylstilbene [98]. Here, more effort needs to be invested to identify the electro-oxidative transformations of other thiophene *S*-oxides.

4.6. Structural studies on thiophene *S*-oxides

In 1990, Rauchfuss et al. published an X-ray crystal structure of the tetramethylthiophene *S*-oxide rhodium complex **56** [34]. The first X-ray single crystal structure determination of a non-liganded thiophene S-oxide was carried out by Meier-Brocks and Weiss on tetraphenylthiophene *S*-oxide. The crystal, however, showed some disorder, and only limited information could be gleaned from it [39]. In 1995, Mansuy et al. carried out an X-ray crystal structural analysis of 2,5-diphenylthiophene *S*-oxide (**160**) [46, 47], where the structure of **160** was compared to 2,5-diphenylthiophene (**162**) and 2,5-diphenylthiophene *S,S*-dioxide (**169**). The S—O bond in the thiophene *S*-oxide was found with 1.484(3) Å to be appreciably longer than those of the thiophene *S,S*-dioxide with 1.418(5) Å and 1.427(5) Å, respectively [47]. The ring system of the thiophene *S,S*-dioxide **169** was found to be absolutely planar, while thiophene *S*-oxide **160** was found to be puckered, with the sulfur lying outside the plane constructed by the four ring carbons by 0.278 Å, and the sulfoxy oxygen lying outside of the plane on the side opposite to sulfur, located by 0.746 Å away from the plane. Previously, this non-planarity of thiophene *S*-oxides had been predicted by MNDO [99] and *ab-initio* calculations [100] of the parent thiophene *S*-oxide itself and dibenzothiophene *S*-oxide. A more pronounced alteration between double and single C—C bond was found in thiophene *S*-oxide **160** in comparison to diphenylthiophene [47]. In probing the aromaticity of thiophene *S*-oxide **160**, it can be seen that apart from its non-planarity, it exhibits relatively large bond order alternations [C(2)—C(3) 2.11; C(3)—C(4) 1.23, C(2)/C(5)—S 1.11; for comparison, the bond orders in 162: C(2)—C(3) 1.94; C(3)—C(4) 1.46; C(2)/C(5) 1.53]. The corresponding 2,5-diphenylthiophene *S,S*-dioxide, though features even larger bond alternations than **160** [47]. An approach for an assessment of aromaticity is the A index as defined by Julg and François [101], which evaluates aromaticity in respect to bond alternation and bond delocalization in ring systems. Here, benzene as the aromatic system par excellence, has an A index of 1, the thiophene system in 2,5-diphenylthiophene has an A index of 0.99, the 5-membered ring system in 2,5-diphenylthiophene *S*-oxide's A index is calculated at 0.79, and the parent thiophene *S*-oxide A index lies at 0.69 ([47], see also [102]).

Subsequently, further X-ray crystal structure analyses were carried out on thiophene S-oxide, such as on 2,5-bis(diphenylmethylsilyl)thiophene *S*-oxide [45], 3,4-bis-*tert*-butylthiophene *S*-oxide (**73**) [43], (1,1,7,7-tetraethyl-3,3,5,5-tetramethyl-*s*-hydrindacen-4-yl)thiophene *S*-oxide [68], 1,3-bis(thien-2yl)-4,5,6,7-tetrahydrobenzo[*c*]thiophene *S*-oxide [40], and the sexithiophene (**170**) (**Figure 7**), where two of the thienyl units were oxidized to sulfoxides [103]. As the thiophene S-oxides are not planar, they invert at the sulfur with different substituents at the C2/C5 positions leading to different barriers of inversion, which have in part been determined experimentally [32, 68, 69]. Structural features of thiophene *S*-oxides and thiophene *S,S*-dioxides have been reviewed before [104].

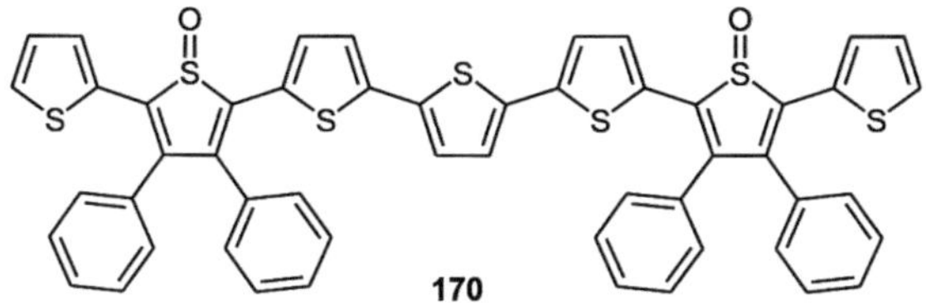
170

Figure 7. Oligothiophene *S*-oxide **170**.

4.7. Oligomers and polymers incorporating thiophene *S*-oxide units

Oligothiophenes and polythiophenes are being studied as advanced materials with interesting electronic and nonlinear optical properties [105] with applications in photovoltaic cells [106] and field effect transistors (FETs) [107], among others. It has been noted that oxidation of thienyl-units in oligothiophenes and polythiophenes leads to a lowering of energy gaps, to greater electron affinities, and to greater ionization energies [103, 108, 109]. The introduction of thienyl-*S*,*S*-dioxides into oligothiophenes often leads to solubility problems of the materials and often leads to a noticeable increase of oxidation potentials. Therefore, there has been a recent interest in incorporating thienyl *S*-oxide units in oligo- and polythiophenes with the aim of greater solubility and smaller oxidation potentials and narrower energy gaps with electron-affinities similar to thienyl *S*,*S*-dioxides [103].

A number of synthetic approaches exist towards the preparation of oligothiophenes with thienyl *S*-oxide units. Oxidation of a pre-prepared oligo- or polythiophene is more difficult to achieve and leads to modest yield [110]. However, two strategies can be seen as promising. One is the transformation of polyarylene-alkynes **171** via oligozirconacyclopentadienes **172** to polythiophene *S*-oxides **173**, where the zirconacyclopentadienes are reacted with SO_2 [41] (**Scheme 44**). The other takes advantage of the fact that certain thiophene *S*-oxides such as 2-bromo-3,4-diphenyl-thiophene *S*-oxide (**175**) are stable enough to be subjected to C—C cross-coupling reactions and subsequent halogenation reactions with *N*-bromosuccinimide (NBS), leading to sequences as shown in **Scheme 45** [103]. Already, an FET has been synthesized with a thienyl-thienyl *S*-oxide polymer [103]. Also, larger π-conjugated ring systems with a thienyl *S*-oxide unit such as **179** have attracted some attention because of their electronic and optical properties (**Figure 8**) [111]. As a drawback, it may be noted that thienyl *S*-oxides in oligomers and polymers would not be stable towards UV radiation as opposed to thienyl *S*,*S*-dioxides [112, 113].

4.8. Thiophene *S*-oxides as metabolites in the enzymatic oxidation of thiophenes

Thiophenes have been known to have toxic effects [114, 115]. The understanding of the mechanism leading to the toxicity of thiophenes is of importance, as a number of drugs such as tienilic acid (**180**), ticlopidine (**182**), methapyrilene (**183**), thenalidine (**184**), tenoxicam (**185**), cephaloridine (**186**), suprofen (**187**), and clopidogrel (**188**) carry thienyl units, where some of the drugs have been taken off the market (**Figure 9**). Already in 1990, it was shown that hepatic cytochrome P450 mediated oxidation of the thienyl-containing tienilic acid (**180**) led to

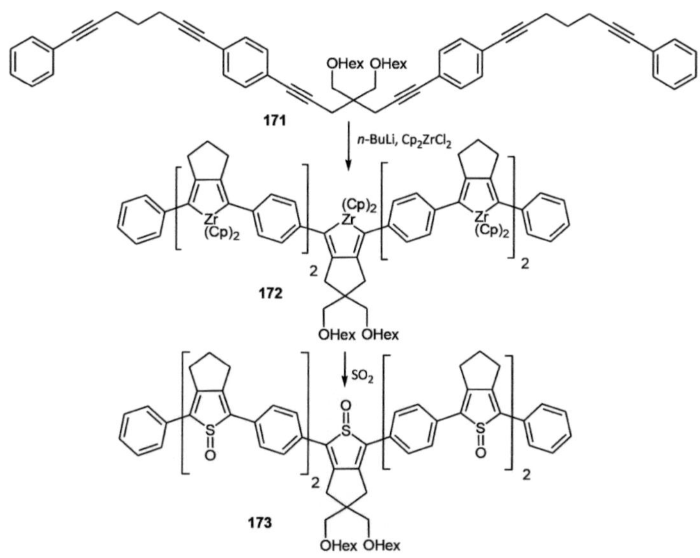

Scheme 44. Preparation with oligomer **173** via zirconacyclopentadiene **172**.

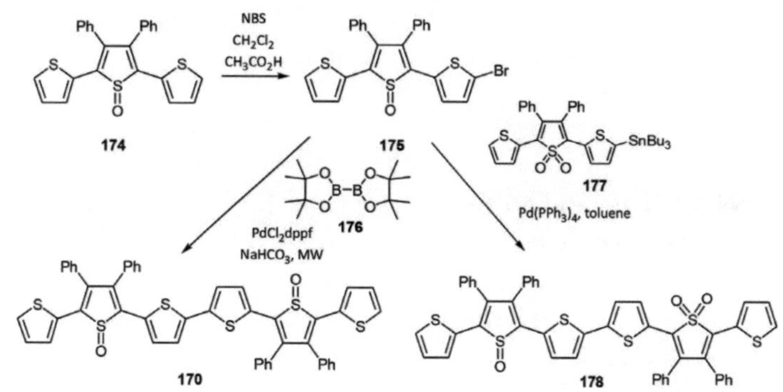

Scheme 45. Preparation of thienyl S-oxide containing oligomers **170** and **178** by Pd(0) Suzuki and Stille cross-coupling reactions.

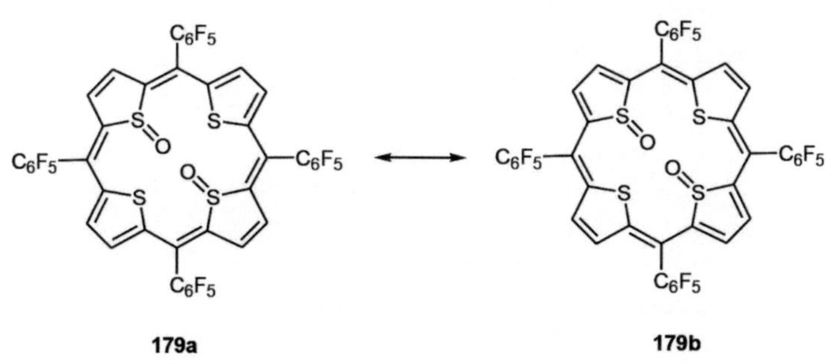

Figure 8. Tetrakis(pentafluorophenyl)tetrathiaisophlorin dioxide (**179**).

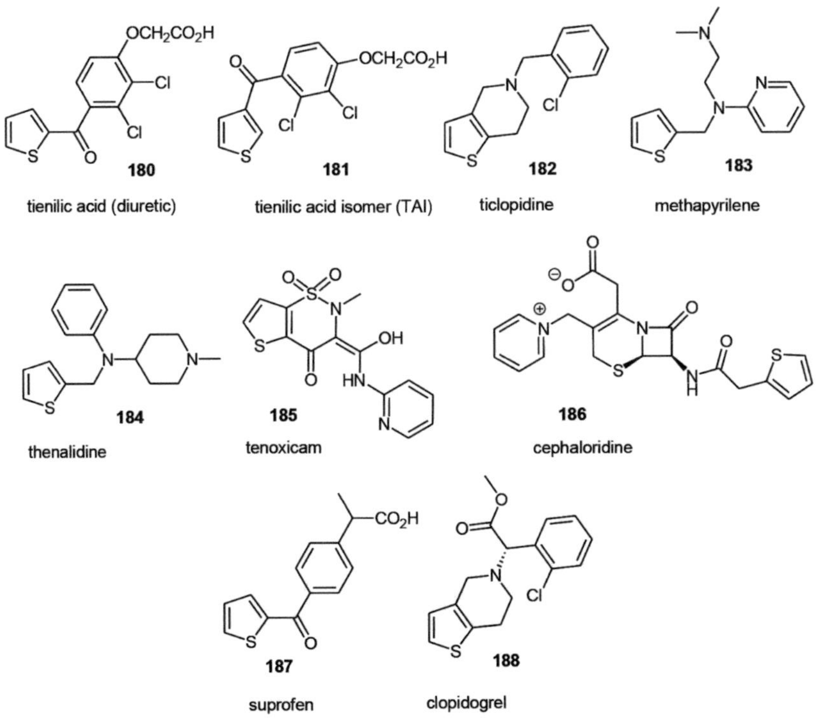

Figure 9. Thiophene-containing pharmaceuticals.

electrophilic metabolites that bind to hepatic proteins [116, 117]. Oxidative metabolism of thiophenes in rats involves thiophene *S*-oxides [118–120]. It has been found [119, 121] that rats administered with thiophene (**1**) in corn oil showed dihydrothiophene *S*-oxide **191** in their urine as a major metabolite [119] (**Scheme 46**). This metabolite was assumed to stem from the addition of glutathione (**189**) to a reactive intermediate thiophene *S*-oxide **3** (**Scheme 46**). Previously, it had been shown that rat liver microsomal cytochrome P450 oxidizes 3-aroylthiophene **181**, a regioisomer of tienilic acid (**180**), to aroylthiophene *S*-oxide **192**, which in the presence of mercaptoethanol (**193**) transformed into dihydrothiophene *S*-oxide **194** [121] (**Scheme 47**). Also, **181** was oxidized by clofibrate induced rat liver microsomes to *S*-oxide **191**,

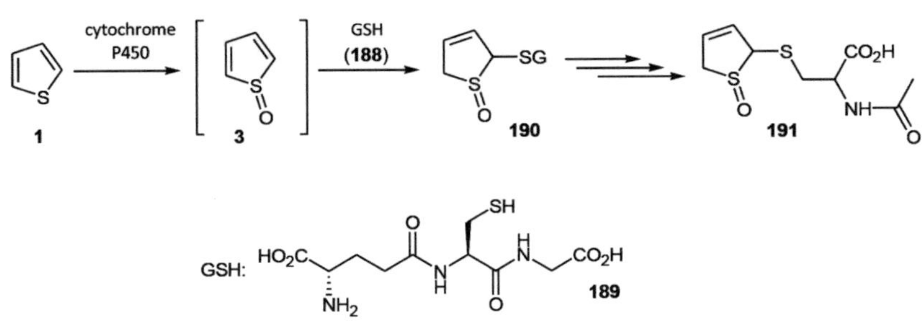

Scheme 46. Cytochrome P450 mediated transformation of thiophene **1** to adduct **191**.

http://dx.doi.org/10.5772/intechopen.79080

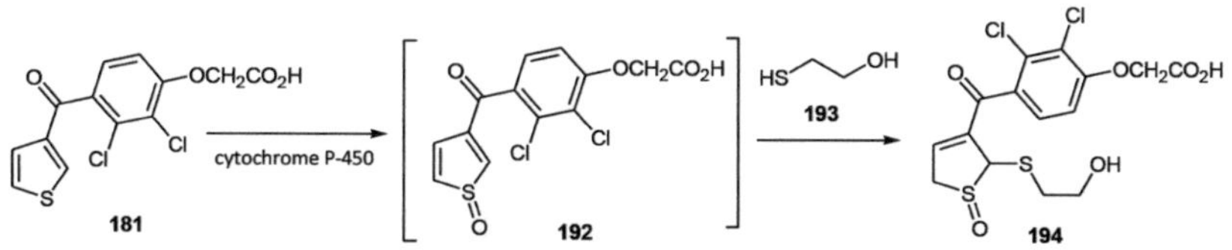

Scheme 47. Transformation of tienilic acid regioisomer **181** to thiophene *S*-oxide and its addition of mercaptoethanol (**193**).

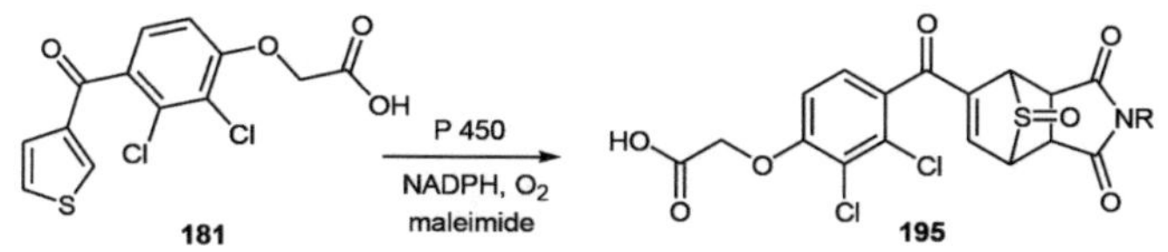

Scheme 48. Cycloaddition of the thiophene S-oxide derivative of **181** to maleimide.

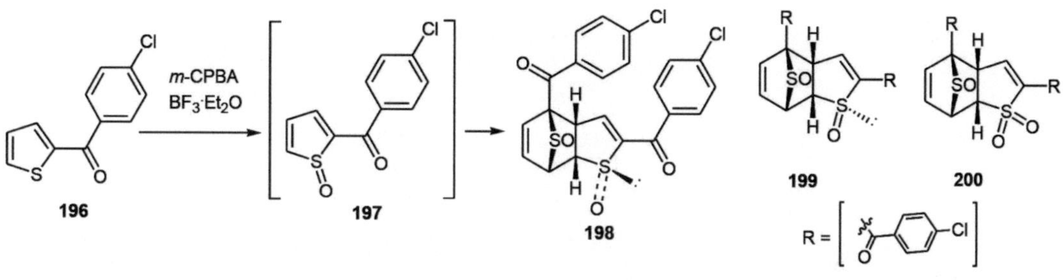

Scheme 49. Formation of sequioxides **198–200** by dimerization of thiophene *S*-oxide **197**.

which was then trapped as a Diels Alder product with maleimides, for example as **195** [120] (**Scheme 48**).

The oxidation of 2-(4-chlorobenzoyl)thiophene (**196**), a molecule in structure close to tienilic acid, by H_2O_2 in the presence of trifluoroacetic acid (TFA) and by *m*-CPBA, $BF_3{\cdot}Et_2O$, both in CH_2Cl_2, gives sesquioxides **198–200** that clearly indicate that a thiophene *S*-oxide structure **197** is formed as an intermediate [122] (**Scheme 49**). Nevertheless, the oxidation of thiophene (**1**) itself with H_2O_2 in the presence of TFA produces apart from sesquioxides **6–8** thiophen-2-one (thiolactone **202**). Thiophen-2-one (**202**) most likely is produced through thiophene-epoxide (**201**) [23] (**Scheme 50**). Thiophen-2-one (**202**) is in equilibrium with 2-hydroxythiophene (**202**). There is one report of a Pummerer-like rearrangement reaction that leads from the purified and isolated thiophene *S*-oxide **134** to thiophen-2-one (thiolactone **202**) [123] (**Scheme 51**). Still, the current understanding is that the thiophene *S*-oxide intermediates formed *in vivo* do not lead to a 2-hydroxythiophene (**203**) [124] (**Scheme 52**), so that two separate mechanisms may exist for the cytochrome P450 2C9 (CYP2C9) mediated oxidation of thiophenes. In this regard, Dansette et al. [119] showed that CYP450s may catalyze both the reaction of thiophenes to thiophene *S*-oxide and to thiophene epoxides [125].

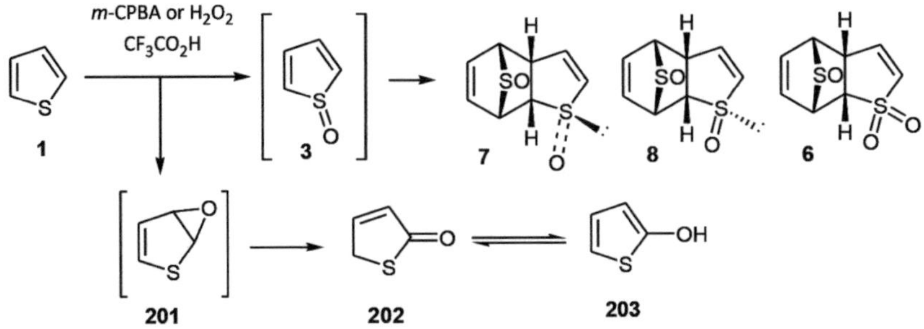

Scheme 50. Reaction of thiophene (**1**) leads via thiophene *S*-oxide (**3**) to sesquioxides **7–9** and in a separate pathway via thiophene epoxide **201** to thiolactone **202** and thus to 2-hydroxythiophene (**203**).

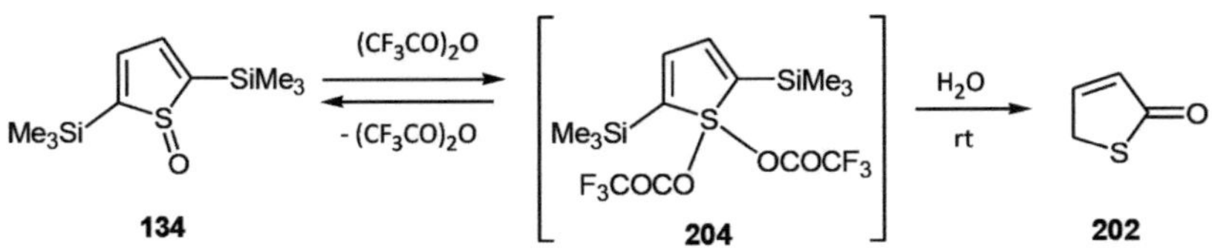

Scheme 51. Pummerer reaction of thiophene *S*-oxide **134** to thiolactone **202**.

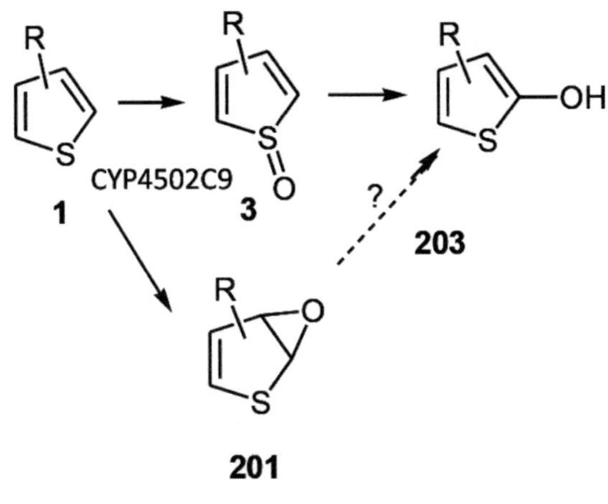

Scheme 52. Cytochrome P450 mediated oxidation of thiophene may lead to two pathways, one through thiophene *S*-oxide **3**, the other through thiophene epoxide **201**.

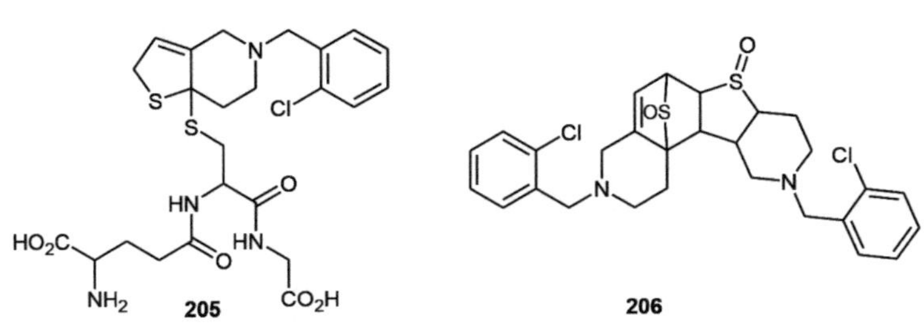

Figure 10. Metabolites of ticlopidine that derive from a ticlopidine *S*-oxide intermediate.

http://dx.doi.org/10.5772/intechopen.79080

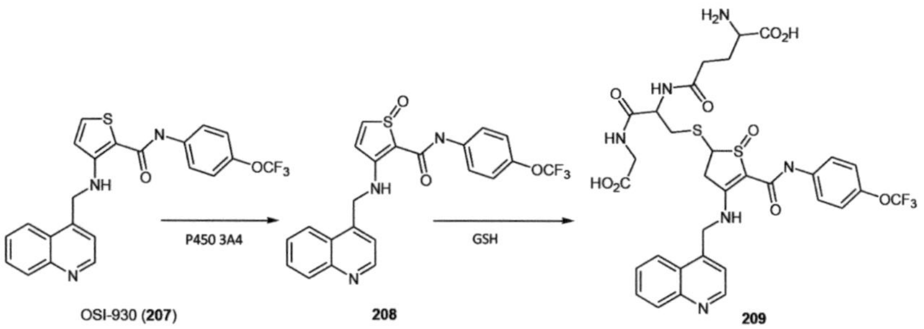

Scheme 53. *In vivo* oxidation of anticancer drug OSI-930 (**207**) to OSI-930 sulfoxide (**208**) and addition of glutathione (GSH) to provide identified metabolite **209**.

Also, the investigation of the metabolism of other thienyl-containing pharmaceuticals show that potentially both mechanisms, epoxidation of the thiophene-unit and oxidation of the thiophene-unit to thiophene *S*-oxide, operate concurrently. As to the thiophene *S*-oxide pathway, Shimizu et al. in their investigation of metabolites ticlopidine (**182**) in rats found both the glutathione conjugate of ticlopidine *S*-oxide **205** and the dimeric ticlopidine *S*-oxide cycloadduct **206** (**Figure 10**) [126, 127]. The structures could be identified by mass spectrometry, and ^{1}H and ^{13}C NMR spectrometry. Medower et al. have noted that cytochrome P450 mediated oxidation of cancer drug OSI-930 (**207**) leads to GSH conjugate **209**, derived from OSI-930 *S*-oxide (**208**), as recognized by mass spectrometry (**Scheme 53**) [128].

Lastly, both possible metabolic pathways of thiophenes, via thiophene *S*-oxides and via thiophene epoxides, have been examined as to their energy profiles using density functional theory [129]. It was found that the formation of the thiophene epoxide (−23.24 kcal/mol) is more exothermic than the formation of the thiophene *S*-oxide (−8.08 kcal/mol) [129]. Also, the formation of thiophene epoxide seems kinetically favored [129]. Both possible metabolites, thiophene *S*-oxide and thiophene epoxide, are highly electrophilic, leading to bond formation with nucleophiles such as with amino acids, leading to a mechanism-based inactivation (MBI) of cytochrome P450.

5. Conclusion

Since the first unverified isolation of a thiophene *S*-oxide a little more than 50 years ago, research on thiophene *S*-oxides has reached a milestone. Due to mainly two synthetic routes, the controlled oxidation of thiophenes in presence of a Lewis- or proton acid and the reaction of zirconacyclopentadienes with thionyl chloride, a number of thiophene *S*-oxides have now become readily accessible. Thiophene *S*-oxides are noted to be reactive dienes in Diels-Alder type cycloadditions, where they react equally well with electron-poor and electron-rich dienophiles. Thiophene *S*-oxides can be stabilized by sterically exacting substituents. Then, they exhibit sufficient stability to be submitted to Pd(0)-catalyzed cross-coupling reactions without deoxygenation.

This leads to the possibility of preparing aryl-oligomers with thiophene-*S*-oxide subunits. By comparing oligothiophenes and oligomers with thiophene *S,S*-dioxide subunits, oligomers with thiophene *S*-oxide subunits exhibit smaller oxidation potentials and narrower energy gaps with electron-affinities greater than oligothiophenes and similar to thiophene *S,S*-dioxides. Nevertheless, thiophene S-oxides are not stable photochemically, but deoxygenate to the corresponding thiophenes or transform to furans by photochemical rearrangement.

Thiophene *S*-oxides have been found to act as intermediates in the cytochrome P540 mediated, oxidative metabolism of thiophene-containing compounds, including a number of important thiophene containing pharmaceuticals. Addition of nucleophiles *in vivo* leads to mechanism based inhibition (MBI) and to toxic side effects of the thiophenes, including nephrotoxicity.

Author details

Thies Thiemann

Address all correspondence to: thies@uaeu.ac.ae

Department of Chemistry, College of Science, United Arab Emirates University, Al Ain, Abu Dhabi, United Arab Emirates

References

[1] Lanfry M. Sur les oxythiophènes. Comptes Rendus. 1911;**153**:73-76

[2] Lanfry M. ur les oxy-b-méthylthiophènes. Comptes Rendus. 1911;**153**:821-822

[3] Bailey WJ, Cummins EW. Cyclic dienes. III. Synthesis of thiophene 1-dioxide. Journal of the American Chemical Society. 1954;**76**:1932-1936

[4] Bailey WJ, Cummins EW. Cyclic dienes. IV. The dimerization of thiophene 1-dioxide. Journal of the American Chemical Society. 1954;**76**:1936-1940

[5] Nagasawa H, Sugihara Y, Ishii A, Nakayama J. Thiophene 1,1-dioxide: Synthesis, isolation, and properties. Bulletin of the Chemical Society of Japan. 1999;**72**:1919-1926

[6] Benders PH, Reinhoudt DN, Trompenaars WP. Thiophene and its derivatives: Part 1. In: Gronowitz S, editor. The Chemistry of Heterocyclic Compounds: A Series of Monographs. Vol. 44. New York: John Wiley & Sons; 1985. p. 713

[7] Benders PH, Reinhoudt DN, Trompenaars WP. Cycloaddition reactions of thiophenes, thiophene 1-oxides, and 1,1-dioxides. Chemistry of Heterocyclic Compounds. 2008;**44**:671-744

[8] Raasch MS. Thiophene-1,1-dioxides, sesquioxides, and 1-oxides. Chemistry of Heterocyclic Compounds. 2008;**44**:571-628

[9] Melles J, Backer HJ. Sesquioxydes obtenus par oxydation de thiophenes: (propriétés du groupe sulfonyle XXXVI). Recueil des Travaux Chimiques des Pays-Bas. 1953;**72**:491-496

[10] van Tilborg WJM. Improved method for the synthesis of dialkyl-substituted thiophene 1,1-dioxides. Synthetic Communications. 1976;**6**:583-589

[11] Li YQ, Thiemann T, Sawada T, Mataka S, Tashiro M. Lewis acid catalysis in the oxidative cycloaddition of thiophenes. The Journal of Organic Chemistry. 1997;**62**:7926-7936

[12] Li YQ, Matsuda M, Thiemann T, Sawada T, Mataka S, Tashiro M. Lewis acid catalyzed oxidative cycloaddition of thiophenes. Synlett. 1996:461-464

[13] Torssell K. Diels-Alder reactions of thiophene oxides generated *in situ*. Acta Chemica Scandinavica. 1976;**30**:353-357

[14] Naperstkow AM, Macaulay JB, Newlands MJ, Fallis AG. Pi-facial diastereoselectivity in Diels-Alder reactions of 2,5-dimethylthiophene oxide. Tetrahedron Letters. 1989;**30**:5077-5080

[15] Li YQ, Thiemann T, Sawada T, Tashiro M. Novel crown ethers by oxidative cycloaddition of thiopheno-crown ethers. Journal of the Chemical Society, Perkin Transactions. 1994;**1**: 2323-2329

[16] Thiemann T, Dongol KG, Thiemann T. A new route to non-natural aryl-containing amino acids and their precursors from thiophenes. Journal of Chemical Research. 2003:527-528; Journal of Chemical Research (S). 2003;**27**:901-942

[17] Thiemann T, Li YQ, Thiemann C, Sawada T, Ohira D, Tashiro M, Mataka S. Oxidative cycloaddition of structures with multiple core thiophenes. Heterocycles. 2000;**52**:1215-1230

[18] Thiemann T, Sa e Melo ML, Campos Neves AS, Li YQ, Mataka S, Tashiro M, Geissler U, Walton DJ. Preparation and electrooxidative SO-extrusion of halogenated 7-thiabicyclo [2.2.1]heptane 7-oxides. Journal of chemical research. Synopses. 1998;**22**:346-346

[19] Arima K, Ohira D, Mataka S, Thiemann T. Cycloaddition reactions of benzo[*b*]thiophene *S*-oxides. IOSR Journal of Applied Chemistry (IOSR-JAC). 2018;**11**(3):19-27

[20] For the parent compound, thiophene *S,S*-dioxide, B. Jursic calculated the HOMO energy to be −11.03 eV (by AM1 semiempirical method): B. Jursic, suitability of furan, pyrrole and thiophene s dienes for Diels-Alder reactions viewed through their stability and reaction barriers for reactions with acetylene, ethylene and cyclopropene. An AM1 semi-empirical and B3LYP hybrid density functional theory study. Journal of Molecular Structure: Theochem. 1998;**454**:105-116

[21] A notable exception is the reactivity tetrahalothiophene S,S-dioxides, which already undergoes cycloadditions at room temperature: Raasch MS. Annulations with tetrachlorothiophene 1,1-dioxides. The Journal of Organic Chemistry. 1980;**45**:856-867

[22] Certain other electron-acceptor substituted thiophene *S,S*-dioxides such as sulfonyl-, alkoxycarbonyl-, or cyano-thiophene *S,S*-dioxides react as dienophiles at room temperature:

Moiseev A, Tyurin DD, Balenkova ES, Nenajdenko VG. Reactions of acceptor substituted thiophene-1,1-dioxides with cyclopentadiene: Control of selectivity by substitution. Tetrahedron. 2006;**62**:4139-4145

[23] Levandowski BJ, Herath D, Gallup NM, Houk KN. Origin of π-selectivity in thiophene 1-oxide cycloadditions. The Journal of Organic Chemistry. 2018;**83**:2611-2616

[24] Thiemann T, Li YQ, Mataka S, Tashiro M. Intramolecular oxidative cycloaddition of thiophenes. Journal of Chemical Research. 1995;**19**:384-384; (M), 1995;**19**:2364-2379

[25] Li YQ, Thiemann T, Mimura K, Sawada T, Mataka S, Tashiro M. Oxidative cycloaddition of thiophenophanes—[n](2.5)parathiophenophane (n=8,10-12,14), [8](2,4)metathiophenophane and [2.2](2,5)parametathiophenophane. European Journal of Organic Chemistry. 1998:1841-1850

[26] Wang Z, Li M, Liu X, Yu B. Synthesis of steroidal saponins bearing an aromatic E ring. Tetrahedron Letters. 2007;**48**:7323-7326

[27] Treiber A, Dansette PM, Mansuy D. Mechanism of the aromatic hydroxylation of thiophene by acid-catalyzed peracid oxidation. The Journal of Organic Chemistry. 2002;**67**: 7261-7266

[28] Wang P, Min J, Nwachukwu JC, Cavett V, Carlson KE, Guo P, Zhu M, Zhang Y, Dong C, Katzenellenbogen JA, Nettles KW, Zhou H-B. Identification and structure—Activity relationships of a novel series of estrogen receptor ligands based on 7-thiabicyclo[2.2.1] hept-2-ene 7-oxde. Journal of Medicinal Chemistry. 2012;**55**:2324-2341

[29] Zeng HP, Eguchi S. Selective formation of mono-and tetra-cycloadducts by the Diels-Alder reaction of [60]fullerene with 2,5-dimethylthiophene S-oxides. Synlett. 1997:175-176

[30] Eguchi S, Ohno M, Kojima S, Shirakawa Y, Zeng HP. [4+2]-cycloaddition of C60 with electron-deficient dienes and application of the adducts to further functionalization in fullerenes, recent advances in the chemistry and physics of fullerenes and related materials (K. M. Kadish, R. S. Ruoff, eds.). The Electrochemical Society Proceedings. 1997; **97**(14):338-346

[31] Thiemann T, Ohira D, Li YQ, Sawada T, Taniguchi M, Tashiro M, Mataka S. Thieno[3.3] orthocyclophanes. Preparation and structures. New Journal of Chemistry. 1999;**23**:675-678

[32] Mock WL. Stable thiophene sulfoxides. Journal of the American Chemical Society. 1970; **92**:7610-7612

[33] Procházka M. Űber Thiophen-1-oxid. Collection of Czechoslovak Chemical Communications. 1965;**30**:1158-1168

[34] Skaugset AE, Rauchfuss TB, Stern CL. Oxidation of coordinated thiophene: Preparation of Cp*Rh(tetramethylthiophene-S-oxide). Journal of the American Chemical Society. 1990;**112**:2432-2433

[35] Krautscheid H, Feng Q, Rauchfuss TB. Base hydrolysis of ruthenium(II) thiophene complexes and reactions of the coordinated ligands. Organometallics. 1993;**12**:3273-3281

[36] Skaugset AE. PhD [thesis]. University of Illinois at Urbana-Champaign; 1992

[37] Fagan PJ, Nugent WA, Calabrese JC. Metallacycle transfer from zirconium to main group elements: A versatile synthesis of heterocycles. Journal of the American Chemical Society. 1994;**116**:1880-1889

[38] Fagan PJ, Nugent WA. Synthesis of main group heterocycles by metallacycle transfer from zirconium. Journal of the American Chemical Society. 1988;**110**:2310-2312

[39] Meier-Brocks F, Weiss E. Tetraphenylzirkonacyclopentadien-derivate als Synthone für Tetraphenylthiophenemonoxid und substituierte Germanole. Journal of Organometallic Chemistry. 1993;**453**:33-45

[40] Jiang B, Tilley TD. General, efficient route to thiophene 1-oxides and well-defined, mixed thiophene-thiophene-1-oxide oligomers. Journal of the American Chemical Society. 1999; **121**:9744-9745

[41] Suh MC, Jiang B, Tilley TD. An efficient, modular synthetic route to oligomers based on zirconocene coupling: Properties for phenylene-thiophene-1,1-dioxide chains. Angewandte Chemie (International Ed. in English). 2000;**39**:2870-2873

[42] Miller RW, Dodge NJ, Dyer AM, Fortner-Buczala EM, Whalley AC. A one-pot method for the preparation of 2,5-diarylthiophene-1-oxides from arylacetylenes. Tetrahedron Letters. 2016;**57**:1860-1862

[43] Nakayama J, Yu T, Sugihara Y, Ishii A. Synthesis and characterization of thiophene 1-oxides kinetically stabilized by bulky substituents at the 3- and 4-positions. Chemistry Letters. 1997;**26**:499-500

[44] Furukawa N, Zhang SZ, Sato S, Higaki M. Simple procedure for the synthesis of 2,5-bis (silylated)thiophene *S*-oxides with *m*-chloroperbenzoic acid in the presence of $BF_3{\cdot}Et_2O$. Heterocycles. 1997;**44**:61-66

[45] Furukawa N, Zhang D, Horn E, Takahashi O, Sato S. X-ray crystallographic analysis of 2,5-bis(diphenylmethylsilyl)thiophene monoxide and the Diels-Alder reaction of thiophene monoxide with dienophiles. Heterocycles. 1998;**47**:793-809

[46] Pouzet P, Erdelmeier I, Ginderow D, Mornon JP, Dansette PM, Mansuy D. Thiophene S-oxides—Convenient preparation, first complete structural characterization and unexpected dimerization of one of them, 2,5-diphenylthiophene-1-oxide. Journal of the Chemical Society, Chemical Communications. 1995;**31**:473-474

[47] Pouzet P, Erdelmeier I, Ginderow D, Mornon JP, Dansette PM, Mansuy D. Thiophene 1-oxides. 5. Comparison of the crystal structures and thiophene ring aromaticity of 2,5-diphenylthiophene, its sulfoxide and sulfone. Journal of Heterocyclic Chemistry. 1997;**34**: 1567-1574

[48] Magerramov AM, Nagieva IT, Allakhverdiev MA. Oxidation of 3,4-dibromo-2,5-dimethylthiophene by hydrogen peroxide on a biomimetic catalyst – Fe(III) ethylenediaminetetraacetate/Al_2O_3. Azerbaidzhanskii Khimicheskii Zhurnal. 2002:27-36

[49] Nagieva IT. The mechanism of the heterogeneous catalytic monooxidation of thiophene derivatives at the S position by hydrogen peroxide. Russian Journal of Physical Chemistry A. 2009;**83**:1873-1878

[50] Ma B, Zhao W, Zhang F, Zhang Y, Wu S, Ding Y. A new halide-free efficient reaction-controlled phase-transfer catalyst based on silicotungstate of $[(C_{18}H_{37})_2(CH_3)_2N]_3[SiO_4H(WO_5)_3]$ for olefin epoxidation, oxidation of sulfides and alcohols with hydrogen peroxide. RSC Advances. 2014;**4**:32054-32062

[51] Thiemann C, Thiemann T, Li YQ, Sawada T, Nagano Y, Tashiro M. SO-photoextrusion of 7-thiabicyclo[2.2.1]hept-2-ene 7-oxides. Bulletin of the Chemical Society of Japan. 1994; **67**:1886-1893

[52] Li YQ, Thiemann C, Ohira D, Mataka S, Tashiro M, Thiemann T. From thiophene S-oxides to 7-thiabicyclo[2.2.1]hept-5-enes. Journal of Chemical Research. 2009;**33**:702-704

[53] Takayama J, Sugihara Y, Takanayagi T, Nakayama J. Syn-pi-face and endo-selective, inverse electron-demand Diels-Alder reactions of 3,4-di-tert-butylthiophene 1-oxide with electron-rich dienophiles. Tetrahedron Letters. 2005;**46**:4165-4169

[54] Otani T, Takayama J, Sugihara Y, Ishii A, Nakayama J. Pi-facial selective Diels-Alder reactions of 3,4-di-tert-butylthiophene 1-oxide and 1-imide and formation of 1,2-thiazetidines. Journal of the American Chemical Society. 2003;**125**:8255-8263

[55] Thiemann T, Ohira D, Li YQ, Sawada T, Mataka S, Rauch K, Noltemeyer M, de Meijere A. [4+2]-cycloaddition of thiophene S-monoxides to activated methylenecyclopropanes. Journal of the Chemical Society, Perkin Transactions. 2000;**1**:2968-2976

[56] Thiemann T, Fujii H, Ohira D, Arima K, Li YQ, Mataka S. Cycloaddition of thiophene *S*-oxides to allenes, alkynes and to benzyne. New Journal of Chemistry. 2003;**27**:1377-1384

[57] Takayama J, Sugihara Y, Nakayama J. SO_2-extrusion of an 8-thiabicyclo[3.2.1]octa-2,6-diene 8,8-dioxide and rearrangement of the resulting cycloheptatriene. Heteroatom Chemistry. 2005;**16**:132-137

[58] Thiemann T, Iniesta J, Walton DJ. [4+2]-cycloaddition of sterically hindered thiophene S-oxides to alkenes and SO-extrusion reactions of the cycloadducts. Phosphorus, Sulfur and Silicon. 2017;**191**:876-884

[59] Iniesta J, Matsumoto T, Thiemann. Cycloaddition of benzo[*b*]thiophene-*S*,*S*-dioxide—A route to substituted dibenzothiophenes and dibenzothiophene *S*,*S*-dioxides. Journal of Chemical Research. 2008;**23**:109-114

[60] Thiemann T, Iniesta J, Walton D. Thermal oxidation of tetracyclones (2,3,4,5-tetraarylcyclopentadienones). Journal of Chemical Research. 2008;**23**:173-180

[61] Iguchi K, Sugihara Y, Nakayama J. Preparation of sterically congested compounds: 6,7-di-t-butyl-1,4-naphthoquinone, 2,3,6,7-tetra-t-butylanthraquinone, and 2,3,6,7-tetra-t-butylanthracene. Bulletin of the Chemical Society of Japan. 2008;**81**:304-306

[62] Thiemann T, Tanaka Y, Iniesta J. Brominated thiophenes as precursors in the preparation of brominated and arylated anthraquinones. Molecules. 2009;**12**:1013-1031

[63] Thiemann T, Tanaka Y, Iniesta J, Varghese HT, Panicker CY. Arylation of chloroanthraquinones by surprisingly facile Suzuki-Miyaura cross-coupling reactions. Journal of Chemical Research. 2009;**33**:732-736

[64] Joseph T, Varghese HT, Panicker CY, Thiemann T, Viswanathan K, van Alsenoy C. Infrared and Raman spectroscopic analysis and theoretical computation of the first hyperpolarizability of a monoarylated anthraquinone, 1-(4-methoxy-phenyl)-4-methylanthraquinone. Journal of Molecular Structure. 2011;**1005**:17-24

[65] Takeuchi N, Nakahodo T, Fujihara H. Remarkably stable *S*-oxides of calix[4]thiophenes and their sulfonium ylide from reaction of *S*-oxide with acetylene derivative. Chemistry Letters. 2017;**46**:389-391

[66] Takayama J, Fukuda S, Sugihara Y, Ishii A, Nakayama J. Pi-facial selective hetero Diels-Alder reactions of 3,4-di-tert-butylthiophene 1-oxide. An excellent trapping reagent for thioaldehydes and thioketones. Tetrahedron Letters. 2003;**44**:5159-5162

[67] Nakayama J, Tai A, Isawa S, Furuya T, Sugihara Y. Tetracyanoethylene oxide not only oxidizes sulfides to sulfoxides but also reduces sulfoxides to sulfides. Tetrahedron Letters. 2005;**46**:1395-1397

[68] Otani T, Miyoshi M, Shibata T, Matsuo T, Hashizume D, Tamao K. Thermally stable monosubstituted thiophene 1-oxide and 1-imides stabilized by a bulky rind group at their 3-position: Synthesis, structure, and inversion barriers on the sulfur atom. Bulletin of the Chemical Society of Japan. 2017;**90**:697-705

[69] Boyd D, Sharma DD, Haughey SA, Malone JF, McMurray BT, Sheldrake GN, Allen CC, Dalton HJ. Enantioselective dioxygenase-catalysed formation and thermal racemization of chiral thiophene sulfoxides. Journal of the Chemical Society, Chemical Communications. 1996;**32**:2363-2364

[70] Houk KN, Stozier RW. Lewis acid catalysis of Diels Alder reactions. Journal of the American Chemical Society. 1973;**95**:4094-4096

[71] Alston PV, Ottenbrite RM. Secondary orbital interactions determining regioselectivity in the Lewis acid catalyzed Diels-Alder reaction II. Journal of Organic Chemistry. 1975;**40**: 1111-1116

[72] Ishida M, Inagaki S. In: Inagaki S, editor. Orbitals in Chemistry. Vol. 289. Berlin: Springer-Verlag; 2009. pp. 183-183

[73] Cieplak AS. Stereochemistry of nucleophilic addition to cyclohexanone. The importance of two-electron stabilizing interactions. Journal of the American Chemical Society. 1981; **103**:4540-4552

[74] Cieplak AS, Tait BD, Johnson CR. Reversal of π-facial diastereoselection upon electronegative substitution of the substrate and the reagent. Journal of the American Chemical Society. 1989;**111**:8447-8462

[75] Cieplak AS. Inductive and resonance effects of substituents on π-face selection. Chemical Reviews. 1999;**99**:1265-1336

[76] Macaulay JB, Fallis AG. Heteroatom-directed pi-facial diastereoselection in Diels-Alder cycloadditions of plane-nonsymmetric cyclopentadienes. Journal of the American Chemical Society. 1990;**112**:1136-1144

[77] Li YQ. Dissertation. Fukuoka, Japan: Kyushu University; 1997

[78] Nakayama J, Takayama J, Sugihara Y, Ishii A. Reaction of 3,4-di-t-butylthiophene 1-oxide with 2-methylene-1,3-dimethylimidazolidine: Methylene transfer and [4+4]-dimerization. Chemistry Letters. 2001;**30**:758-759

[79] Takayama J, Sugihara A, Ishii A, Nakayama J. Preparation of 1,4,5,8-tetra-*tert*-butyl-1,3,5,7-cyclooctatetraene by twofold SO_2 extrusion. Tetrahedron Letters. 2003;**44**:7893-7896

[80] Nakayama J, Furuya T, Suzuki Y. Pi-face-selective 1,3-dipolar cycloadditions of 3,4-di-tert-butylthiophene 1-oxide with 1,3-dipoles. Phosphorus, Sulfur and Silicon. 2009;**184**: 1175-1183

[81] Nakayama J, Furuya T, Ishii A, Sakamoto A, Otani T, Sugihara Y. Regio- and stereoselective addition of molecular bromine to S-oxidized derivatives of 3,4-di-t-butylthiophene: Exclusive 1,4-cis additions. Bulletin of the Chemical Society of Japan. 2003;**76**:619-625

[82] Zhang S-Z, Sato S, Horn E, Furukawa N. Exclusive cis-1,4-addition in the bromination of 2,5-bis(trimethylsilyl)-thiophene monoxide. Heterocycles. 1998;**4**:227-234

[83] Otani T, Sugihara Y, Ishii A, Nakayama J. Michael additions of oxygen and sulfur nucleophiles to 3,4-di-t-butyl-1[(p-tolyl)sulfonylimino]-1,1-dihydrothiophene. A comparison study with 3,4-di-t-butylthiophene 1-oxide and 1,1-dioxide. Chemistry Letters. 2000;**29**:744-745

[84] Nakayama J, Aoki S, Takayama J, Sakamoto A, Sugihara Y, Ishii A. Reversible disulfur monoxide (S_2O)-forming retro-Diels-Alder reaction. Disproportionation of S_2O to trithioozone (S_3) and sulfur dioxide (SO_2) and reactivities of S_2O and S_3. Journal of the American Chemical Society. 2004;**126**:9085-9093

[85] Otani T, Sugihara Y, Ishii A, Nakayama J. Preparation of 3,4-di-t-butylthiophene 1-imide and its N-substituted derivatives. Tetrahedron Letters. 2000;**41**:8461-8465

[86] Nakayama J, Otani T, Sugihara Y, Sano Y, Ishii A, Sakamoto A. Syntheses and structures of sulfilimine, sulfone diamine, and sulfoximine derivatives of a monocyclic thiophene, 3,4-di-tert-butylthiophene. Heteroatom Chemistry. 2001;**12**:333-348

[87] Nakayama J, Sano Y, Sugihara Y, Ishii A. Thiophene sulfoximides: 2,4- and 3,4-di-tert-butyl-1-imino-1,1-dihydrothiophene 1-oxides. Tetrahedron Letters. 1999;**40**:3785-3788

[88] Gurria GM, Posner GH. Photochemical deoxygenation of aryl sulfoxides. The Journal of Organic Chemistry. 1973;**38**:2419-2420

[89] Wan Z, Jenks WS. Oxenoid reactivity observed on the photolysis of certain aromatic sulfoxides. Journal of the American Chemical Society. 1995;**117**:2667-2668

[90] Gregory DD, Wan Z, Jenks WS. Photodeoxygenation of dibenzothiophene sulfoxide: Evidence for a unimolecular S—O cleavage mechanism. Journal of the American Chemical Society. 1997;**119**:94-102

[91] Rockafellow EM, McCulla RD, Jenks WS. Deoxygenation of dibenzothiophene S-oxide and dibenzoselenophene S-oxide: A comparison of direct and sensitized photolysis. Journal of Photochemistry and Photobiology A. 2008;**198**:45-51

[92] Arima K, Ohira D, Watanabe M, Miura A, Mataka S, Thiemann T, Iniesta J, Walton D. The photochemistry of thiophene S-oxides. Photochemical & Photobiological Sciences. 2005;**4**:808-816

[93] Heying MJ, Nag M, Jenks WS. Photochemistry of thiophene S-oxide derivatives. Journal of Physical Organic Chemistry. 2008;**21**:915-924

[94] Stoffregen SA, Lee SY, Dickerson P, Jenks WS. Computational investigation of the photochemical deoxygenation of thiophene-S-oxide and selenophene-Se-oxide. Photochemical & Photobiological Sciences. 2014;**13**:431-438

[95] Nakayama J, Hiraiwa S, Fujihara T. Photolysis and photo-oxidation of 3,4-di-tert-butylthiophene 1-oxide. Journal of Sulfur Chemistry. 2008;**29**:243-250

[96] Thiemann T, Ohira D, Arima K, Sawada T, Mataka S, Marken F, Compton RG, Bull SD, Davies SG. Photochemical and electrochemical behavior of thiophene S-oxides. Journal of Physical Organic Chemistry. 2000;**13**:648-653

[97] Valcarel JI, Walton DJ, Fujii H, Thiemann T, Tanaka Y, Mataka S, Mason TJ, Lorimer JP. The sonoelectrooxidation of thiophene S-oxides. Ultrasonics Sonochemistry. 2004;**11**:227-232

[98] Iniesta J, Alcock H, Walton DJ, Watanabe M, Mataka S, Thiemann T. Electrochemical oxidation of tetracyclones and tetraphenylthiophene S-oxide. Electrochimica Acta. 2006; **51**:5682-5690

[99] Hashmall JA, Horak V, Khoo LE, Quicksall CO, Sun MK. Molecular structure of selected *S*-methylthiophenium tetrafluoroborates and dibenzothiophene *S*-oxide. Journal of the American Chemical Society. 1981;**103**:289-295

[100] Rozas I. Comparative study of aromaticity in five-membered rings containing S, SO, and SO_2 groups. Journal of Physical Organic Chemistry. 1992;**5**:74-82

[101] Julg A, François P. Recherches sur la géométrie de quelques hydrocarbures non-alternants: Son influence sur les énergies de transition, une nouvelle définition de l'aromaticité. Theoretica Chimica Acta. 1967;**7**:249-259

[102] Holloczki O, Nyulaszi L. Analogy between sulfuryl and phosphino groups: The aromaticity of thiophene-oxide. Structural Chemistry. 2011;**22**:1385-1392

[103] Di Maria F, Zangoli M, Palama IE, Fabiano E, Zanelli A, Monari M, Perinot A, Caironi M, Maiorano V, Maggiore A, Pugliese M, Salatelli E, Gigli G, Viola I, Barbarella G. Improving the property-function tuning range of thiophene materials via facile synthesis of oligo/polythiophene-S-oxides/oligo/polythiophene-S,S-dioxides. Advanced Functional Materials. 2016;**26**:6970-6984

[104] Lukevics E, Arsenyan P, Belyakov S, Pudova O. Molecular structure of thiophene 1,1-dioxides, thiophene S-oxides and their derivatives. Chemistry of Heterocyclic Compounds. 2002;**38**:632-645

[105] Fichou D, editor. Handbook of Oligo- and Polythiophenes. Wiley-VCH Verlag; 1999

[106] Malytskyi V, Simon JJ, Patrone L, Raimundo JM. Thiophene-based push-pull chromophores for small molecule organic solar cells (SMOCs). RSC Advances. 2015;**5**:354-397

[107] Takimiya K, Osaka I, Mori T, Nakano M. Organic semiconductors based on [1]benzothieno [3,2-b][1]benzothiophene substructure. Accounts of Chemical Research. 2014;**47**:1493-1502

[108] Tanaka K, Wang S, Yamabe T. Electronic structures of substituted derivatives of polythiophene. Design of narrow-band-gap polymers. Synthetic Metals. 1989;**30**:57-65

[109] Bongini A, Barbarella G, Zambianchi M, Arbizzani C, Mastragostino M. Thiophene S-oxides: Orbital energies and electrochemical properties. Journal of the Chemical Society, Chemical Communications. 2000:439-440

[110] Sano R, Suzuki T, Kawai T, Iyoda T. Direct sulfur oxidation of poly(3-hexylthiophene). Memoirs of the Graduate School of Engineering, Tokyo Metropolitan University. 1999;**49**: 183-193

[111] Mishra VL, Furuyama T, Kobayashi N, Goto K, Miyazaki T, Yang J-S, Shinmyozu T. Synthesis, optical properties, and electronic structures of tetrakis(pentafluorophenyl) tetrathiaisophlorin dioxide. Chemistry—A European Journal. 2016;**22**:9190-9197

[112] Thiemann T, Kumazoe K, Arima K, Mataka S. Studies towards dibenzothiophene S-oxide arrays and their photochemical activity. Rep. Inst. Adv. Mat. Kyushu Univ. 2001; **15**(1):63-71; Chemical Abstracts. 2002;**136**:19-802

[113] Kumazoe K, Arima K, Mataka S, Walton DJ, Thiemann T. Photobehaviour of substituted dibenzothiophene-S-oxides and oxygenated bis(dibenzothiophene)s. Journal of Chemical Research. 2003;**27**(S):60-61; 2003;**27**(M):248-264

[114] McMurty RJ, Mitchell JR. Renal and hepatic necrosis after metabolic activation of 2-substituted furans and thiophenes including furosemide and cephaloridine. Toxicology and Applied Pharmacology. 1977;**42**:285-300

[115] Beaune P, Dansette PM, Mansuy D, Kiffel L, Finck M, Amar C, Leroux JP, Homberg JC. Human anti-endoplasmatic reticulum autoantibodies appearing in a drug-induced hepatitis are directed against a human liver cytochrome P-450 that hydroxylates the drug. Proceedings of the National Academy of Sciences of the United States of America. 1987; **84**:551-555

[116] Dansette PM, Amar C, Smith C, Pons C, Mansuy D. Oxidative activation of the thiophene ring by hepatic enzymes. Hydroxylation and formation of electrophilic metabolites during metabolism of tienilic acid and its isomer by rat liver microsomes. Biochemical Pharmacology. 1990;**39**:911-918

[117] Dansette PM, Amar C, Valadon P, Pons C, Beaune PH, Mansuy D. Hydroxylation and formation of electrophilic metabolites of tienilic acid and its isomer by human liver microsomes. Catalysis by a cytochrome P450 IIC different from that responsible for mephenytoin hydroxylation. Biochemical Pharmacology. 1991;**40**:553-560

[118] Treiber A, Dansette PM, El Amri H, Girault J-P, Ginderow D, Mornon J-P, Mansuy D. Chemical and biological oxidation of thiophene: Preparation and complete characterization of thiophene S-oxide dimers and evidence for thiophene S-oxide as an intermediate in thiophene metabolism in vivo and in vitro. Journal of the American Chemical Society. 1997;**119**:1565-1571

[119] Dansette PM, Thang DC, El Amri H, Mansuy D. Evidence for thiophene S-oxide as a primary reactive metabolite of thiophene in vivo: Formation of a dihydrothiophene sulfoxide mercapturic acid. Biochemical and Biophysical Research Communications. 1992;**186**:1624-1630

[120] Dansette PM, Thebault S, Durand-Gasselin L, Mansuy D. Reactive metabolites of thiophenic compounds: A new trapping method for thiophene sulfoxides. Drug Metabolism Reviews. 2010;**32**(Suppl. 1):233-234 presentation at ISSX, Istanbul

[121] Mansuy D, Valadon P, Erdelmeier I, Lopez-Garcia P, Amar C, Girault JP, Dansette PM. Thiophene S-oxides as new reactive metabolites: Formation by cytochrome P450 dependent oxidation and reaction with nucleophiles. Journal of the American Chemical Society. 1991;**113**:7825-7826

[122] Ho MT, Treiber A, Dansette PM. Oxidation of 2-(4-chlorobenzoyl)thiophene into 1-oxide Diels-Alder dimers, sesquioxide and a sulfone-water adduct. Tetrahedron Letters. 1998; **39**:5049-5052

[123] Sato S, Zhang S-Z, Furukawa N. The Pummerer-like reaction of 2,5-bis(trimethylsilyl) thiophene S-oxide with trifluoroacetic anhydride: Intermediary formation of sulfurane. Heteroatom Chemistry. 2001;**15**:444-450

[124] Rademacher PM, Woods CM, Huang QB, Szklarz GD, Nelson SD. Differential oxidation of two thiophene-containing regioisomers to reactive metabolites by cytochrome P450 2C9. Chemical Research in Toxicology. 2012;**25**:895-903

[125] Dansette PM, Bertho G, Mansuy D. First evidence that cytochrome P450 may catalyze both S-oxidation and epoxidation of thiophene derivatives. Biochemical and Biophysical Research Communications. 2005;**338**:450-455

[126] Shimizu S, Atsumi R, Nakazawa T, Fujimaki Y, Sudo K, Okazaki O. Metabolism of ticlopidine in rats: Identification of the main biliary metabolite as a glutathione conjugate of ticlopidine S-oxide. Drug Metabolism and Disposition. 2009;**37**:1904-1915

[127] Ha-Duong NT, Dijols S, Macherey AC, Goldstein JA, Dansette PM, Mansuy D. Ticlopidine as a selective mechanism-based inhibitor of human cytochrome P450 2C19. Biochemistry. 2001;**40**:12112-12122

[128] Medower C, Wen L, Johnson WW. Cytochrome P450 oxidation of the thiophene-containing anticancer drug 3-[(quinolin-4-ylmethyl)-amino]-thiophene-2-carboxylic acid (4-trifluoromethoxyphenyl)amide to an electrophilic intermediate. Chemical Research in Toxicology. 2008;**21**:1570-1577

[129] Jaladanski CK, Taxak N, Varikoti RA, Bharatam PV. Toxicity originating from thiophene containing drugs: Exploring the mechanism using quantum chemical methods. Chemical Research in Toxicology. 2015;**28**:2364-2376

Chapter 5

The Role of Sulfur-Related Species in Oxygen Reduction Reactions

Dan Xu and Winston Duo Wu

Additional information is available at the end of the chapter

http://dx.doi.org/10.5772/intechopen.78647

Abstract

Heteroatom (metal and nonmetal) doping is essential to achieve excellent oxygen reduction reaction (ORR) activity of carbon materials. Among the heteroatoms that have been studied to date, sulfur (S) doping, including metal sulfides and sulfur atoms, has attracted tremendous attention. Since S-doping can modify spin density distributions around the metal centers as well as the synergistic effect between S and other doped heteroatoms, the S-C bond and metal sulfides can function as important ORR active sites. Furthermore, the S-doped hybrid sample shows a small charge-transfer resistance. Therefore, S-doping contributes to the superior ORR performance. This chapter describes the recent advancements of S-doped carbon materials, and their development in the area of ORR with regard to components, structures, and their ORR activities of S-related species.

Keywords: S-doping, metal sulfides, sulfur atoms, oxygen reduction reactions, active sites

1. Introduction

Fuel cells are considered as promising energy conversion and storage devices. In such a device, fuels (such as hydrogen, methanol, ethanol, or formic acid) react with oxygen at the anode, while oxygen molecules are reduced to water molecules at the cathode [1–4]. However, the oxygen reduction reaction (ORR) rate is ~5 orders of magnitude slower than the reaction on the anode due to its high overpotential [5]. The search for catalysts that can conquer these huge activation energy barriers has attracted much attention. Although Pt-based electrocatalysts have been commercialized, the high cost of Pt and their poor tolerance to methanol significantly hamper their large-scale commercialization. Thus, great effort has been devoted to developing low cost, non-precious-metal, and metal-free catalysts with improved electrocatalytic efficiency [6–9].

Excellent electrocatalysts for ORR should possess a high specific surface area, finely tuned pore structure, and good electron conductivity. The former two facilitate easy accessibility to the active sites and ion diffusion, and the latter is beneficial for electron transfer. Much attention has been focused on the carbonaceous materials due to their remarkable advantages, such as low cost, facile preparation strategy, and high conductivity. For constructing ORR catalysts with promising electrocatalytic activity, single atom doping or co-doping of two or multiple heteroatoms are essential. Metal/nitrogen/carbon (M/N/C) catalysts have been regarded as the most promising alternative for precious metal catalysts. For example, Fe species not only facilitate the formation of catalytically active N-C sites, but Fe atoms also contribute to the graphitization of carbon. More importantly, Fe atoms and related nanoparticles are generally suggested as the active site of ORR catalysts. Recently, the introduction of nonmetal heteroatoms such as N, P, S, or B into carbon materials is generally effective in enhancing ORR activities of catalysts. In N-doped carbon, the N atom with higher electronegativity (3.04) than that of carbon (2.55) leads to more charged adjacent C atoms. With respect to S, the electronegativity of S (2.58) is similar to that of carbon; however, S can easily change the band gap of carbon due to its two lone pair electrons [2]. P with an electronegativity of 2.19 and B with an electronegativity of 2.04 can also induce imbalanced charge distribution in carbon materials, thus forming positively polarized C-P and C-B more active sites to ORR [10, 11]. Furthermore, N/B, N/P, N/S, and N/S/P co-doped carbons also show excellent catalytic activity due to their synergistic effects on spin or charge density of carbon matrix. Notably, designing a carbon matrix with different morphologies combined with hierarchical porous structures, such as micro-, meso-, and macroporosity, can further optimize ORR activity.

Recently, the S atom has attracted particular interest and especially its high synergetic effect with N dopants and metal dopants [12–19]. In this chapter, we briefly summarize the S-related species as active sites in the ORR, such as S-M/N/C, metal chalcogenides, N/S, and N/S/P. We then discuss the S-containing electrocatalysts including their carbon sources, heteroatom dopants, and preparation methods as well as the nanostructure of the supports.

2. S-related active sites in the ORR

The development of novel strategies for the design and synthesis of carbon-based high performance electrocatalysts is a hot topic. Therefore, efforts have been made toward the design and synthesis of extraordinary ORR catalytic carbon materials with different morphologies and single or multiple active sites from diverse sources. S-related active sites have been extensively investigated due to their excellent performance in ORR. S atom mono-doping can induce structural defects in the carbon matrix. The resulting charge dislocation can improve the oxygen adsorption. Furthermore, protonation of S is not as severe as that of N [20]. More importantly, dual doping of N and S or multiple doping of N, S, and M [21] can dramatically enhance ORR activity due to the synergistic effects.

2.1. M-N-S-based active sites

Wu et al. prepared Fe, N, and S decorated hierarchical carbon layers (S, N-Fe/N/C-CNT) from pyrolysis of 2,2-bipyridine and $Fe(SCN)_3$-coated CNTs [22]. Adding S salts not only contributed to the formation of atomically dispersed Fe-N_x species, but also improved the surface area of the carbon matrix. The half-wave potential ($E_{1/2}$) of the S,N-Fe/N/C-CNT catalyst is about 0.85 V, which is higher than that of commercial Pt/C (0.82 V). The catalyst also exhibited superior durability in alkaline medium. Theoretical calculations predicted that atomically dispersed Fe-N_x species function as highly active sites, while co-doping of N and S improved the electrical conductivity. Furthermore, Wan et al. fabricated a sandwich-like graphene/carbon hybrid from graphene oxide (GO) and nontoxic starch (**Figure 1**) [23]. Graphene/carbon nanosheets decorated by N, S, and Fe (Fe, S/NGC) were obtained via treatment with $FeCl_3$ and KSCN. Fe,S/NGC showed outstanding ORR performance in alkaline medium ($E_{1/2}$ of 0.83 V vs. RHE, surpassing $E_{1/2}$ of NGC (0.76 V) and the Pt/C catalyst (0.81 V)), due to the simultaneous introduction of Fe and S. The Fe_3N and S were considered major active centers in this hybrid. Furthermore, Fe,S/NGC also displayed a high ORR activity in the acidic solution. In addition, an S and N dual-doped Fe-N-S electrocatalyst (Fe-M-LA/C) was obtained via pyrolysis of the mixture of melamine, lipoic acid, $FeCl_3$, and carbon black [24]. FeS and Fe_3C formed in the Fe-M-LA/C. It has been suggested that Fe^{2+} has high catalytic activity in ORR and that Fe_3C is the active site for the ORR. Combined with the N and S-doping, Fe-M-LA/C showed promising ORR activity. Interestingly, sewage sludge itself can be used as "all-in-one" precursor for ORR catalysts [25]. The innate N, Fe, and S compounds in the sewage sludge function as N, Fe, and S dopants. The N, Fe, and S self-doped nanoporous carbon material exhibited favorable electrocatalytic activity in both alkaline and acidic environments. The nanostructure of the carbon matrix also played an important role in ORR. Wan et al. synthesized Fe/N/S-doped carbon from glucose, thiourea, and iron nitrate based on a dual-template method. Multiple active sites such as graphitic-N, pyridinic-N, thiophene-S, FeN_x,

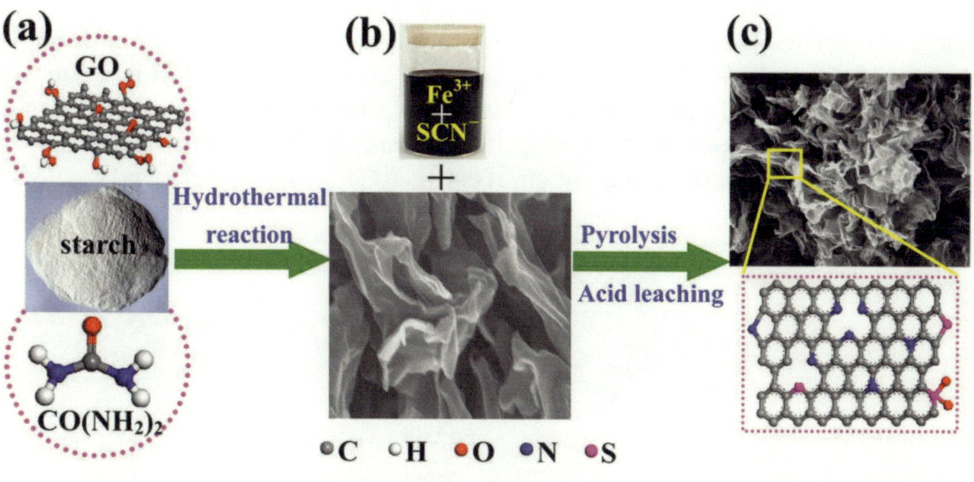

Figure 1. (a) The raw materials of synthesis of NGC nanosheets used as the precursor of Fe,S/NGC-900. (b) Mixed aqueous solution of $FeCl_3$ and KSCN (above) and NGC nanosheets prepared by hydrothermal reaction (below). (c) The as-obtained catalyst (above) and illustration of nitrogen and sulfur atoms in carbon skeleton (below) of Fe,S/NGC-900.

and encapsulated iron nanoparticles combined with hierarchical porous structures contributed to the excellent ORR performance [2].

Bimetal-based N and S doped catalysts have also been reported. Li et al. synthesized PdW alloy nanoparticles decorated S-doped graphene via a microwave irradiation method [26]. S-doping contributed to the formation of small particles and the uniform distribution of alloy particles. The as-prepared catalyst was highly active for ORR due to the specific electronic structure of the alloy. $CoFe_2O_4$ nanoparticles decorated rGO designed by Yang et al. demonstrated high ORR activity due to the existence of defects resulting from the doping of N and S and the covalent coupling between the $CoFe_2O_4$ and rGO matrix [27]. Moreover, Ren et al. synthesized PdNi decorated N and S co-doped three-dimensional ordered carbon derived from acrylonitrile telomere (C, N, and S sources) using silica as template [28]. Due to the co-doping of N and S, the strong electronic interaction between Pd and Ni, and three-dimensional honeycomb-ordered structure, the electrocatalyst exhibited superior performance compared to commercial Pd/C in alkaline solution.

2.2. Metal chalcogenide-based active sites

Wang et al. prepared a raisin bread-like N and S co-doped mesoporous graphitic carbon spheres with $Fe_{1-x}S$ nanocrystals embedded in ($Fe_{1-x}S$/N, S-MGCS) (**Figure 2**) [29]. The $Fe_{1-x}S$/N, S-MGCS catalyst was obtained via pyrolysis of Fe^{2+}-Polydopamine (PDA), followed by a vulcanization process, in which Fe_xC_y was transformed into $Fe_{1-x}S$. This catalyst showed excellent ORR performance in both alkaline medium and acidic medium. Notably, the corresponding E_{onset} and $E_{1/2}$ of $Fe_{1-x}S$/N and S-MGCS were 0.97 and 0.91 V, respectively. The RHE was superior to that of the commercial Pt/C catalyst with an E_{onset} of 0.93 V and an $E_{1/2}$ of 0.87 V. Similarly, Wang et al. prepared a S-Fe/N/C electrocatalyst by pyrolyzing thiourea and iron acetate [30]. Five types of nanoparticles were detected: Fe, FeS, FeN, FeC, and Fe_3O_4. The catalyst showed higher ORR performance compared to Fe/N/C both in alkaline and acidic media. Apparently, more S doping contributed to the higher catalytic performance.

Cobalt chalcogenides as active sites have also attracted significant attention. For example, Li et al. successfully anchored Co_9S_8 nanoparticles on N and S dual doped carbon nanosheets (Co_9S_8/N,S-CNS) via facile pyrolysis of $CoCl_2$, citric acid, and thiourea as carbon source and cubic NaCl crystals were used as template (**Figure 3**) [31]. Due to the highly dispersed nanoparticle and the synergistic catalytic effect between Co_9S_8 nanoparticles and the doped N

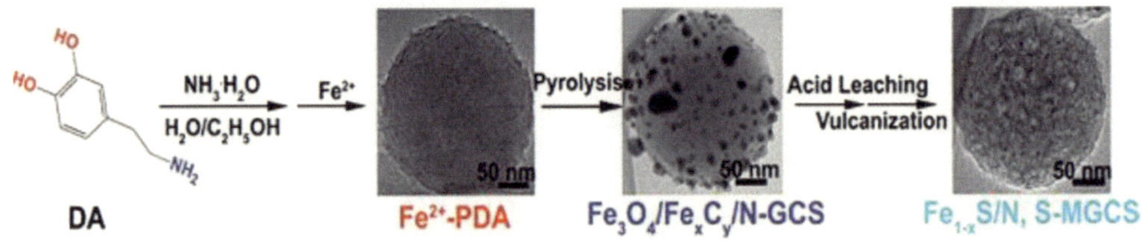

Figure 2. Schematic graph of the formation process from DA to $Fe_{1-x}S$/N, S-MGCS. Reproduced with permission from Ref. [29]. Copyright 2017, Royal Society of Chemistry.

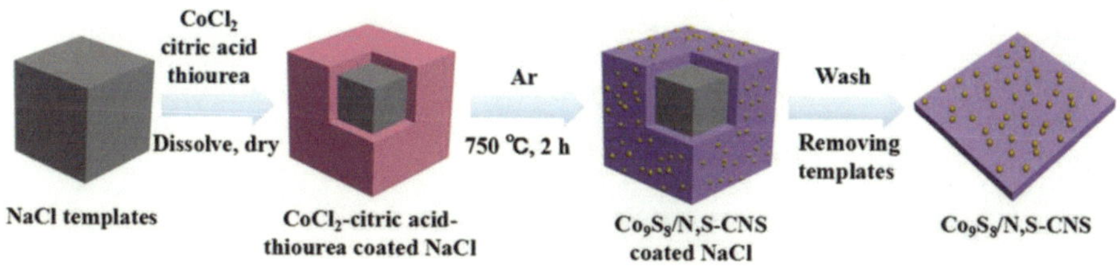

Figure 3. Schematic of the formation of Co_9S_8/N,S-CNS. Reproduced with permission from Ref. [31]. Copyright 2017, Royal Society of Chemistry.

and S in the carbon nanosheets, Co_9S_8/N,S-CNS showed high catalytic activity and stability. Moreover, Liao et al. prepared S co-doped graphene nanoholes with cobalt sulfide hollow nanospheres decorated in ($Co_{1-x}S$/N-S-G) using GO (graphene oxide), phen, and S [32]. The catalyst presented high ORR catalytic activity with an $E_{1/2}$ of 30 mV, which was more positive than that of a commercial Pt/C catalyst. Similarly, Zhang et al. prepared CoS decorated N, S co-doped reduced GO aerogel showing highly efficient activity for ORR [33].

There are other metal sulfides as active sites in ORR. Suh et al. prepared nano-CuS@Cu-BTC composites using Cu-MOF as a sacrificial template and thioacetamide as sulfide source [34]. With the increasing amount of nano-CuS in the composite, electrical conductivity increased, thus contributing to the positive shifts of E_{onset}. MoS_2-embedded nitrogen-doped porous carbon nanosheets were prepared using MoS_2 nanosheets as templates and conjugated microporous polymers as N and C sources [35]. The novel electrocatalysts showed enhanced performance for ORR, due to their strong interaction between MoS_2 and carbon layer, high conductivity, and high specific surface area. Recently, it has been reported that Ni_3S_2 [36] and WS_{3-x} [37] are also potential catalyst for ORR. Furthermore, metallic double sulfides as an ORR catalyst were investigated in recent years. Li et al. prepared $NiCo_2S_4$ and N, S-doped graphene aerogel hybrid for application in ORR [38].

2.3. N-, S-, B-, and P-based active sites

Recently, metal-free catalysts have received much attention, and intensive research efforts have been made. For example, Sun et al. synthesized N,S-co-doped nanocarbon polyhedral morphology using a metal organic framework (MOF) as precursor followed by thermal treatment with ammonia gas (NH_3) and further co-doping of S (**Figure 4**) [39]. The obtained catalyst showed improved electrocatalytic efficiency, comparable to that of the Pt/C catalyst.

Mu et al. synthesized N and S dual-doped 3D porous graphene from waste biomass and GO (**Figure 5**) [40]. The resultant catalyst showed high ORR performance and stability comparable to commercial Pt/C in both alkaline and acidic media due to their unique porous structure and synergistic effects of N and S doping. Furthermore, Wang et al. prepared N and S co-doped 3D hollow structured carbon spheres based on a soft template method [41]. Aniline and pyrrole function as carbon source and N dopant and Na_2S serve as S dopant. The obtained hollow carbon spheres with uniform size, mesoporous structure, and high number of active sites exhibited high ORR activity comparable to that of Pt/C. In contrast, Liao et al.

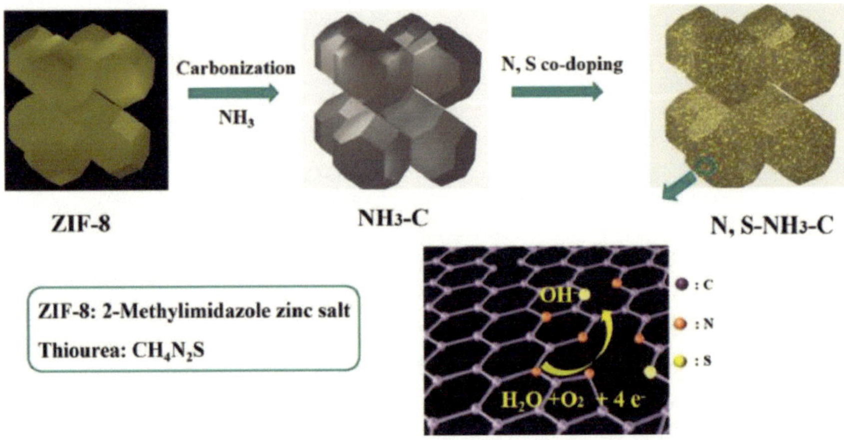

Figure 4. Schematic illustration of the fabrication of the N,S-co-doped nanocarbon as the electrocatalyst toward ORR. The ZIF-8 precursor and thiourea are used as C/N and S precursors, respectively. Reproduced with permission from Ref. [39]. Copyright 2017, Royal Society of Chemistry.

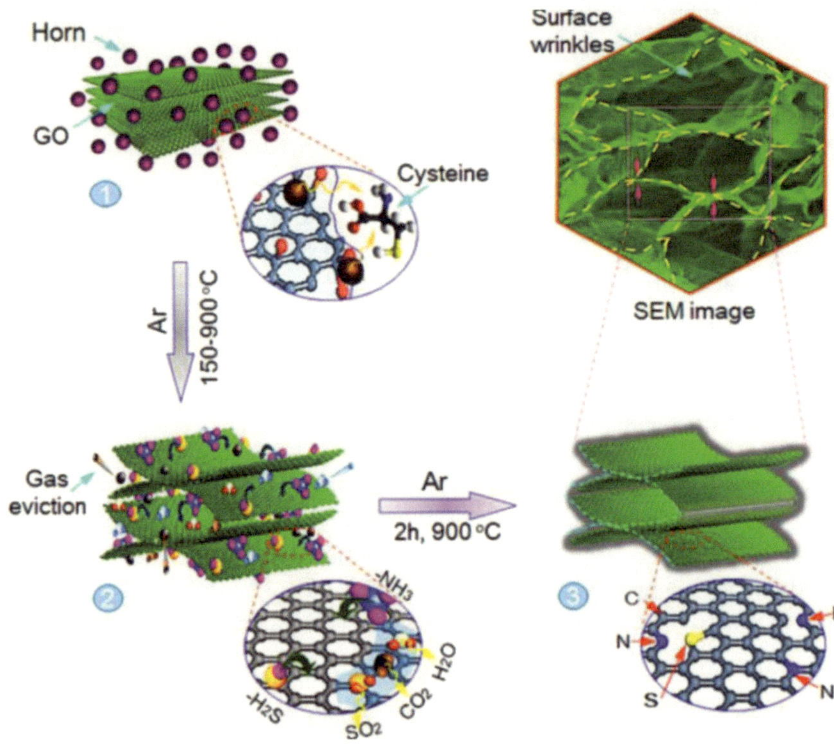

Figure 5. Schematic illustration of the formation of NSG: (stage 1) homogeneous mixture of graphene oxide and horn, (stage 2) disintegration/release of cysteine moieties and coverage of GO surface leading to reaction of functional groups, eviction of gaseous species, and the formation of S and N containing moieties (e.g., H_2S, NH_3, etc.), and (stage 3) doping of N and S into the graphene carbon network. Reproduced with permission from Ref. 40. Copyright 2016, American Chemical Society.

prepared N and S co-doped hollow carbon nanospheres from polyacrylonitrile and S via a hard template method [42]. They reported that S-doping facilitated the formation of pyridinic N, which is more active than other N species in ORR. The catalyst exhibits excellent ORR performance with high stability and selectivity.

Dai et al. reported the development of N, S co-doped graphitic sheets from melamine (as carbon precursor and nitrogen dopant), Ni_2SO_4 (as S dopant and template), and KCl (as template) [43]. The unique hierarchical porous structure renders active sites easily accessible and facilitates electron and mass transfer. Therefore, this catalyst was not only effective in ORR, but also demonstrated excellent activities in OER/HER. In addition, Zhi et al. reported that atomic S doping in mesoporous carbon-supported C_3N_4 can remarkably enhance ORR activity [44]. In this work, thiourea was selected as S dopant and C_3N_4 serve as N source. XPS analysis showed the formation of $C_3N_{4-x}S_x$, indicating the atomic modification over the C_3N_4.

Song et al. prepared S-N dual doped ordered mesoporous carbon based on a hard template method [45]. In their work, polythiophene (PTh) and polypyrrole (PPy) were used as precursors and ordered mesoporous silica (SBA-15) was selected as template. Based on this method, N and S contents can be easily adjusted. Furthermore, the mesoscopic morphology provided more accessible active sites. Therefore, this catalyst showed excellent ORR performance.

Recently, P/S binary-doped carbon materials have also been reported. P, with higher electron-donating ability, heavily affects neighboring carbon atoms, thus tending to induce more active sites than N. For instance, Cao et al. prepared P/S co-doped porous carbon derived from resorcinol, furaldehyde, and phosphorus pentasulfide. Due to the synergetic merits of P and S, the onset potential positively shifted for ORR in alkaline media [46].

2.4. Mechanism

In fact, the mechanism of S-related active sites in ORR is still debated. Suib et al. prepared S-doped carbon nanotube-graphene nanolobes via sequential bidoping strategy, in which the nature of S functionalization can be finely tuned [47]. First, thiourea functioned as the S source. To further stabilize and enhance the content of S, the second doping of benzyl disulfide was introduced. Different doping types of S were detected, such as C-S, C-S-C, and -SO_x-. The S-doped CNT showed high catalytic activity and good stability for ORR. Furthermore, Guo et al. studied the effect of Fe/N/C and C-S-C active sites in alkaline and acidic media [48]. It is worth noting that no Fe-S bond formed in the catalyst. They found that no synergistic effects between Fe/N/C and C-S-C were observed in alkaline solution as the two active centers are separated. In contrast, synergistic effects between Fe/N/C and C-S-C sites remarkably enhanced ORR activity in acidic media because the C-S-C active sites facilitated the 4e- ORR pathway.

Furthermore, S can function as platinum nanowire catalyst anchoring sites. Chen et al. studied the influence of S content on the ORR activity of S-doped graphene supported platinum nanowires (PtNW/SGs) [49]. S doping increased the band gap, while the electrical conductivity decreased. PtNW/SGs with 1.40 at% S showed the best ORR performance. Zhi et al.

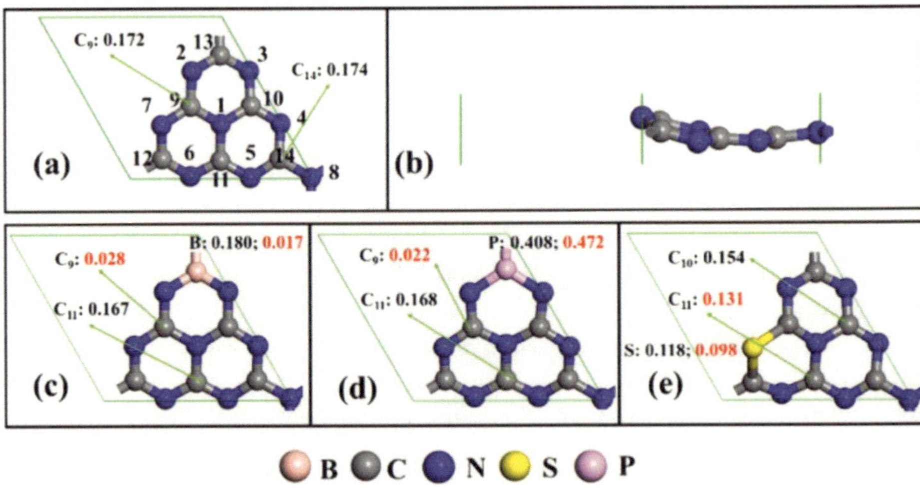

Figure 6. Optimized structure of pristine g-C_3N_4 as (a) top view and (b) side view, in which the N atoms are numbered from 1 to 8, while C atoms are numbered from 9 to 14. Top views of the optimized structures of the energetically most favorable (c) B-CN, (d) P-CN, and (e) SCN, in which the B and P atoms substitute the bay carbon C13, while S atom replaces the pyridinic nitrogen N7. In each structure, the largest value of charge and spin densities on carbon atoms are indicated by black and red colors, respectively; additionally, the related carbon atoms are illustrated by green arrows. Reproduced with permission from Ref. [50]. Copyright 2017, American Chemical Society.

investigated the componential influences of heteroatoms doping (B, P, and S) in graphitic C_3N_4 (g-C_3N_4)-based electrocatalysts (**Figure 6**) [50]. They found that S-doped C_3N_4 with the smallest charge-transfer resistance dramatically boosted the reaction kinetics and activities of ORR.

Recently, Xu et al. designed Fe-N-, Fe-S-, and Fe-N-S-based model catalysts to investigate heteroatom induced performance differences in ORR [51]. Pyrrole-derived and thiophene-derived hypercrosslinked polymers were selected as carbon precursors. $FeCl_3$, a Friedel-Crafts reaction catalyst, acts as both a metal dopant and a porogen. Interestingly, $Fe_{1-x}S$ and Fe_3O_4 nanoparticles formed in the S-doped and N-doped carbon, respectively. In fact, N/Fe_3O_4 acts as a higher catalytic active site than S/$Fe_{1-x}S$. The possible reason is that the strong electronegativity of N generates more charged active sites, while the electronegativity of S is similar to that of carbon. However, the synergistic effect between $Fe_{1-x}S$/Fe_3O_4 and the N, S-doped carbon showed superior ORR performance.

3. Conclusions

Although state-of-the-art Pt-based ORR catalysts are applicable in fuel cell vehicles, source scarcity limits their mass application. M-N-C materials are still far from satisfaction for commercialization. Presently, design and synthesis of novel ORR catalysts with various structures were at the center of research. Furthermore, to experimentally and theoretically explore the relationship between component structure-properties has attracted extensive interest.

http://dx.doi.org/10.5772/intechopen.78647

Particularly, tuning the mode of heteroatom-doping and the underlying the role of active sites in ORR catalysis still remains challenging.

Currently, S-related species represent promising active sites for ORR catalysis. S doping can lead to a higher degree of graphitization because S can react with imperfect carbon to form CS_2 gas [52]. Furthermore, S-doping can modify the spin density distributions around the carbon framework. More importantly, the synergistic effect between the metal center and the N, S-codoped carbon contributes to the superior ORR performance. With regard to metal free catalysts, first-principle calculations indicate that N and S atoms close to each other were more active than isolated N and S sites, indicating a synergistic effect of N and S. Therefore, S-related active sites containing ORR catalyst will be promising alternatives for commercial Pt/C catalysts, especially those with hierarchical porous structures.

Acknowledgements

This work was supported by the National Nature Science Foundation of China (No. 21603156), Jiangsu Province Science Foundation for Youths (No. BK20170331).

Acronyms and abbreviations

ORR	oxygen reduction reaction
PEM	proton exchange membrane
Pt/C	platinum/carbon black catalyst
CB	carbon black
CNT	carbon nanotube
XPS	X-ray photoelectron spectroscopy
$FeCl_3$	iron(III) chloride
KSCN	potassium sulfocyanide
NH_3	ammonia gas
OER	oxygen evolution reaction
HER	hydrogen evolution reaction
rGO	reduced graphene oxide
P	phosphorus
(M/N/C)	metal/nitrogen/carbon

Author details

Dan Xu and Winston Duo Wu*

*Address all correspondence to: duo.wu@suda.edu.cn

Suzhou Key Laboratory of Green Chemical Engineering, School of Chemical and Environmental Engineering, College of Chemistry, Chemical Engineering and Materials Science, Soochow University, Suzhou, China

References

[1] Xia W, Mahmood A, Liang Z, Zou R, Guo S. Earth-abundant nanomaterials for oxygen reduction. Angewandte Chemie, International Edition. 2016;**55**:2650-2676. DOI: 10.1002/anie.201504830

[2] Kone I, Xie A, Tang Y, Chen Y, Liu J, Chen Y, Sun Y, Yang X, Wan P. Hierarchical porous carbon doped with iron/nitrogen/sulfur for efficient oxygen reduction reaction. ACS Applied Materials & Interfaces. 2017;**9**:20963-20973. DOI: 10.1021/acsami.7b02306

[3] Zhang C, An B, Yang L, Wu B, Shi W, Wang YC, Long LS, Wang C, Lin W. Sulfur-doping achieves efficient oxygen reduction in pyrolyzed zeolitic imidazolate frameworks. Journal of Materials Chemistry A. 2016;**4**:4457-4463. DOI: 10.1039/c6ta00768f

[4] Zhu J, Li K, Xiao M, Liu C, Wu Z, Ge J, Xing W. Significantly enhanced oxygen reduction reaction performance of N-doped carbon by heterogeneous sulfur incorporation: Synergistic effect between the two dopants in metal-free catalysts. Journal of Materials Chemistry A. 2016;**4**:7422-7429. DOI: 10.1039/c6ta02419j

[5] Sui S, Wang X, Zhou X, Su Y, Riffat S, Liu CJ. A comprehensive review of Pt electrocatalysts for oxygen reduction reaction: Nanostructure, activity, mechanism and carbon support in PEM fuel cells. Journal of Materials Chemistry A. 2017;**5**:1808-1825. DOI: 10.1039/c6ta08580f

[6] Jin X, Xie Y, Huang J. Highly effective dual transition metal macrocycle based electrocatalyst with macro-/mesoporous structures for oxygen reduction reaction. Catalysts. 2017;**7**:201. DOI: 10.3390/catal7070201

[7] Akhter T, Islam MM, Faisal SN, Haque E, Minett AI, Liu HK, Konstantino K, Dou SX. Self-assembled N/S codoped flexible graphene paper for high performance energy storage and oxygen reduction reaction. ACS Applied Materials & Interfaces. 2016;**8**:2078-2087. DOI: 10.1021/acsami.5b10545

[8] Oh S, Kim JH, Kim M, Nam D, Park JY, Cho EA, Kwon HS. Synergetic effects of edge formation and sulfur doping on the catalytic activity of a graphene-based catalyst for the oxygen reduction reaction. Journal of Materials Chemistry A. 2016;**4**:14400-14407. DOI: 10.1039/c6ta05020d

[9] Bhange SN, Unni SM, Kurungot S. Nitrogen and sulphur co-doped crumbled graphene for the oxygen reduction reaction with improved activity and stability in acidic medium. Journal of Materials Chemistry A. 2016;**4**:6014-6020. DOI: 10.1039/c6ta00073h

[10] Razmjooei F, Singh KP, Yu JS. Superior pore network retention of carbon derived from naturally dried ginkgo leaves and its enhanced oxygen reduction performance. Catalysis Today. 2016;**260**:148-157. DOI: 10.1016/j.cattod.2015.06.012

[11] Tong Y, Chen P, Zhou T, Xu K, Chu W, Wu C, Xie Y. A bifunctional hybrid electrocatalyst for oxygen reduction and evolution: Cobalt oxide nanoparticles strongly coupled to B,N-decorated graphene. Angewandte Chemie, International Edition. 2017;**56**:1-6. DOI: 10.1002/anie.201702430

[12] Wu M, Liu Y, Zhu Y, Lin J, Liu J, Hu H, Wang Y, Zhao Q, Lv R, Qiu J. Supramolecular polymerization-assisted synthesis of nitrogen and sulfur dual-doped porous grapheme networks from petroleum coke as efficient metal-free electrocatalysts for the oxygen reduction reaction. Journal of Materials Chemistry A. 2017;**5**:11331-11339. DOI: 10.1039/c7ta03264a

[13] Li J, Zhang Y, Zhang X, Huang J, Han J, Zhang Z, Han X, Xu P, Song B. S, N dual-doped graphene-like carbon nanosheets as efficient oxygen reduction reaction electrocatalysts. ACS Applied Materials & Interfaces. 2017;**9**:398-405. DOI: 10.1021/acsami.6b12547

[14] Zhao H, Zhu YP, Ge L, Yuan ZY. Nitrogen and sulfur co-doped mesoporous hollow carbon microspheres for highly efficient oxygen reduction electrocatalysts. International Journal of Hydrogen Energy. 2017;**42**:19010-19018. DOI: 10.1016/j.ijhydene.2017.06.172

[15] Yao Y, You Y, Zhang G, Liu J, Sun H, Zou Z, Sun S. Highly functional bioinspired Fe/N/C oxygen reduction reaction catalysts: Structure-regulating oxygen sorption. ACS Applied Materials & Interfaces. 2016;**8**:6464-6471. DOI: 10.1021/acsami.5b11870

[16] Huo L, Liu B, Zhang G, Si R, Liu J, Zhang J. 2D layered non-precious metal mesoporous electrocatalysts for enhanced oxygen reduction reaction. Journal of Materials Chemistry A. 2017;**5**:4868-4878. DOI: 10.1039/c6ta10261a

[17] Liu T, Guo YF, Yan YM, Wang F, Deng C, Rooney D, Sun KN. CoO nanoparticles embedded in three-dimensional nitrogen/sulfur co-doped carbon nanofiber networks as a bifunctional catalyst for oxygen reduction/evolution reactions. Carbon. 2016;**106**:84-92. DOI: 10.1016/j.carbon.2016.05.007

[18] Mulyadi A, Zhang Z, Dutzer M, Liu W, Deng Y. Facile approach for synthesis of doped carbon electrocatalyst from cellulose nanofibrils toward high-performance metal-free oxygen reduction and hydrogen evolution. Nano Energy. 2017;**32**:336-346. DOI: 10.1016/j.nanoen.2016.12.057

[19] Su Y, Yao Z, Zhang F, Wang H, Mics Z, Cánovas E, Bonn M, Zhuang X, Feng X. Sulfur-enriched conjugated polymer nanosheet derived sulfur and nitrogen co-doped porous carbon nanosheets as electrocatalysts for oxygen reduction reaction and zinc-air battery. Advanced Functional Materials. 2016;**26**:5893-5902. DOI: 10.1002/adfm.201602158

[20] Zhang L, Wang Y, Wan K, Piao JH, Liang ZX. Effective sulfur-doping in carbon by high-temperature molten salt bath and its electrocatalysis for oxygen reduction reaction. Electrochemistry Communications. 2018;**86**:53-56. DOI: https://doi.org/10.1016/j.elecom.2017.11.015

[21] Zhu YN, Cao CY, Jiang WJ, Yang SL, Hu JS, Song WG, Wan LJ. Nitrogen, phosphorus and sulfur co-doped ultrathin carbon nanosheets as a metal-free catalyst for selective oxidation of aromatic alkanes and the oxygen reduction reaction. Journal of Materials Chemistry A. 2016;**4**:18470-18477. DOI: 10.1039/c6ta08335h

[22] Chen P, Zhou T, Xing L, Xu TY, Xie H, Zhang L, Yan W, Chu W, Wu C, Xie Y. Atomically dispersed iron–nitrogen species as electrocatalysts for bifunctional oxygen evolution and reduction reactions. Angewandte Chemie, International Edition. 2017;**56**:610-614. DOI: 10.1002/anie.201610119

[23] Men B, Sun Y, Liu J, Tang Y, Chen Y, Wan P, Pan J. Synergistically enhanced electrocatalytic activity of sandwich-like N-doped graphene/carbon nanosheets decorated by Fe and S for oxygen reduction reaction. ACS Applied Materials & Interfaces. 2016;**8**:19533-19541. DOI: 10.1021/acsami.6b06329

[24] Huang HC, Lin YC, Chang ST, Liu CC, Wang KC, Jhong HP, Lee JF, Wang CH. Effect of a sulfur and nitrogen dual-doped Fe-N-S electrocatalyst for the oxygen reduction reaction. Journal of Materials Chemistry A. 2017;**5**:19790-19799. DOI: 10.1039/c7ta05030e

[25] Yuan SJ, Dai XH. Facile synthesis of sewage sludge derived in-situ multi-doped nanoporous carbon material for electrocatalytic oxygen reduction. Scientific Reports. 2016;**6**:27570. DOI: 10.1038/srep27570

[26] Li Y, Li W, Ke T, Zhang P, Ren X, Deng L. Microwave-assisted synthesis of sulfur-doped graphene supported PdW nanoparticles as a high performance electrocatalyst for the oxygen reduction reaction. Electrochemistry Communications. 2016;**69**:68-71. DOI: 10.1016/j.elecom.2016.06.006

[27] Yan W, Cao X, Tian J, Jin C, Ke K, Yang R. Nitrogen/sulfur dual-doped 3D reduced graphene oxide networkssupported $CoFe_2O_4$ with enhanced electrocatalytic activities for oxygen reduction and evolution reactions. Carbon. 2016;**99**:195-202. DOI: 10.1016/j.carbon.2015.12.011

[28] Li Y, Lin S, Ren X, Mi H, Zhang P, Sun L, Deng L, Gao Y. One-step rapid in-situ synthesis of nitrogen and sulfur co-doped three-dimensional honeycomb-ordered carbon supported PdNi nanoparticles as efficient electrocatalyst for oxygen reduction reaction in alkaline solution. Electrochimica Acta. 2017;**253**:445-454. DOI: 10.1016/j.electacta.2017.08.143

[29] Xiao JW, Xia YT, Hu CC, Xi JGB, Wang S. Raisin bread-like iron sulfides/nitrogen and sulfur dual-doped mesoporous graphitic carbon spheres: A promising electrocatalyst for the oxygen reduction reaction in alkaline and acidic media. Journal of Materials Chemistry A. 2017;**5**:11114-11123. DOI: 10.1039/c7ta02096a

[30] Hu K, Tao L, Liu DD, Huo J, Wang SY. Sulfur-doped Fe/N/C Nanosheets as highly efficient electrocatalysts for oxygen reduction reaction. ACS Applied Materials & Interfaces. 2016;**8**:19379-19385. DOI: 10.1021/acsami.6b02078

[31] Wu C, Zhang Y, Dong D, Xie H, Li J. Co_9S_8 nanoparticles anchored on nitrogen and sulfur dual-doped carbon nanosheets as highly efficient bifunctional electrocatalyst for oxygen evolution and reduction reactions. Nanoscale. 2017;**9**:12432-12440. DOI: 10.1039/c7nr03950f

[32] Qiao X, Jin J, Fan H, Li Y, Liao S. In situ growth of cobalt sulfide hollow nanospheres embedded in nitrogen and sulfur co-doped graphene nanoholes as a highly active electrocatalyst for oxygen reduction and evolution. Journal of Materials Chemistry A. 2017;**5**:12354-12360. DOI: 10.1039/c7ta00993c

[33] Luo Z, Tan C, Zhang X, Chen J, Cao X, Li B, Zong Y, Huang L, Huang X, Wang L, Huang W, Zhang H. Preparation of cobalt sulfide nanoparticle-decorated nitrogen and sulfur co-doped reduced graphene oxide aerogel used as a highly efficient electrocatalyst for oxygen reduction reaction. Small. 2016;**12**:5920-5926. DOI: 10.1002/smll.201602615

[34] Cho K, Han SH, Suh MP. Copper–organic framework fabricated with CuS nanoparticles: Synthesis, electrical conductivity, and electrocatalytic activities for oxygen reduction reaction. Angewandte Chemie, International Edition. 2016;**55**:15301-15305. DOI: 10.1002/anie.201607271

[35] Yuan K, Zhuang X, Fu H, Brunklaus G, Forster M, Chen Y, Feng X, Scherf U. Two-dimensional core-shelled porous hybrids as highly efficient catalysts for the oxygen reduction reaction. Angewandte Chemie, International Edition. 2016;**55**:6858-6863. DOI: 10.1002/anie.201600850

[36] Yan B, Concannon NM, Milshtein JD, Brushett FR, Surendranath Y. A membrane-free neutral pH formate fuel cell enabled by a selective nickel sulfide oxygen reduction catalyst. Angewandte Chemie, International Edition. 2017;**56**:1-5. DOI: 10.1002/anie.201702578

[37] Tan SM, Pumera M. Electrosynthesis of bifunctional WS_{3-x}/reduced graphene oxide hybrid for hydrogen evolution reaction and oxygen reduction reaction electrocatalysis. Chemistry—A European Journal. 2017;**23**:1-11. DOI: 10.1002/chem.201701722

[38] Yang T, Li R, Li Z, Gu Z, Wang G, Liu J. Hybrid of $NiCo_2S_4$ and nitrogen and sulphur-functionalized multiple graphene aerogel for application in supercapacitors and oxygen reduction with significant electrochemical synergy. Electrochimica Acta. 2016;**211**:59-70. DOI: 10.1016/j.electacta.2016.06.028

[39] Song Z, Liu W, Cheng N, Banis MN, Li X, Sun Q, Xiao B, Liu Y, Lushington A, Li R, Liu L, Sun X. Origin of the high oxygen reduction reaction of nitrogen and sulfur co-doped MOF-derived nanocarbon electrocatalysts. Materials Horizons. 2017;**4**:900-907. DOI: 10.1039/c7mh00244k

[40] Amiinu IS, Zhang J, Kou Z, Liu X, Asare OK, Zhou H, Cheng K, Zhang H, Mai L, Pan M, Mu S. Self-organized 3D porous graphene dual-doped with biomass-sponsored nitrogen and sulfur for oxygen reduction and evolution. ACS Applied Materials & Interfaces. 2016;**8**:29408-29418. DOI: 10.1021/acsami.6b08719

[41] Wu Z, Liu R, Wang J, Zhu J, Xiao W, Xuan C, Lei W, Wang D. Nitrogen and sulfur co-doping of 3D hollowstructured carbon spheres as an efficient and stable metal free

catalyst for the oxygen reduction reaction. Nanoscale. 2016;**8**:19086-19092. DOI: 10.1039/c6nr06817k

[42] You C, Jiang X, Han L, Wang X, Lin Q, Hua Y, Wang C, Liu X, Liao S. Uniform nitrogen and sulphur co-doped hollow carbon nanospheres as efficient metal-free electrocatalysts for oxygen reduction. Journal of Materials Chemistry A. 2017;**5**:1742-1748. DOI: 10.1039/c6ta08674h

[43] Hu C, Dai L. Multifunctional carbon-based metal-free electrocatalysts for simultaneous oxygen reduction, oxygen evolution, and hydrogen evolution. Advanced Materials. 2017;**29**:1604942. DOI: 10.1002/adma.201604942

[44] Pei Z, Zhao J, Huang Y, Huang Y, Zhu M, Wang Z, Chen Z, Zhi C. Toward enhanced activity of a graphitic carbon nitride-based electrocatalyst in oxygen reduction and hydrogen evolution reactions via atomic sulfur doping. Journal of Materials Chemistry A. 2016;**4**:12205-12211. DOI: 10.1039/c6ta03588d

[45] Jiang T, Wang Y, Wang K, Liang Y, Wu D, Tsiakaras P, Song S. A novel sulfur-nitrogen dual doped ordered mesoporous carbon electrocatalyst for efficient oxygen reduction reaction. Applied Catalysis B: Environmental. 2016;**189**:1-11. DOI: 10.1016/j.apcatb.2016.02.009

[46] Zhou Y, Ma R, Candelaria SL, Wang J, Liu Q, Uchaker E, Li P, Chen Y, Cao G. Phosphorus/sulfur Co-doped porous carbon with enhanced specific capacitance for supercapacitor and improved catalytic activity for oxygen reduction reaction. Journal of Power Sources. 2016;**314**:39-48. DOI: 10.1016/j.jpowsour.2016.03.009

[47] El-Sawy AM, Mosa IM, Su D, Guild CJ, Khalid S, Joesten R, Rusling JF, Suib SL. Controlling the active sites of sulfur-doped carbon nanotube–graphene nanolobes for highly efficient oxygen evolution and reduction catalysis. Advanced Energy Materials. 2016;**6**:1501966. DOI: 10.1002/aenm.201501966

[48] Shen H, Gracia-Espino E, Ma J, Zang K, Luo J, Wang L, Gao S, Mamat X, Hu G, Wagberg T, Guo S. Synergistic effects between atomically dispersed Fe-N-C and C-S-C for the oxygen reduction reaction in acidic media. Angewandte Chemie, International Edition. 2017;**56**:13800-13804. DOI: 10.1002/anie.201706602

[49] Hoque MA, Hassan FM, Seo MH, Choi JY, Pritzker M, Knights S, Ye S, Chen Z. Optimization of sulfur-doped graphene as an emerging platinum nanowires support for oxygen reduction reaction. Nano Energy. 2016;**19**:27-38. DOI: 10.1016/j.nanoen.2015.11.004

[50] Pei ZX, Gu JX, Wang YK, Tang ZJ, Liu ZX, Huang Y, Huang Y, Zhao JX, Chen ZF, Zhi CY. Component matters: Paving the roadmap toward enhanced electrocatalytic performance of graphitic C_3N_4-based catalysts via atomic tuning. ACS Nano. 2017;**11**:6004-6014. DOI: 10.1021/acsnano.7b01908

[51] Zhang J, Xu D, Wang C, Guo J, Yan F. Rational design of $Fe_{1-x}S/Fe_3O_4$/nitrogen and sulfur-doped porous carbon with enhanced oxygen reduction reaction catalytic activity. Advanced Materials Interfaces. 2018:1701641. DOI: 10.1002/admi.201701641

[52] Shih PT, Dong RX, Shen SY, Vittal R, Lin JJ, Ho KC. Transparent graphene–platinum nanohybrid films for counter electrodes in high efficiency dye-sensitized solar cells. Journal of Materials Chemistry A. 2014;**2**:8742-8748. DOI: 10.1039/c3ta12931d

Vibrational Study and Crystal Structure of Barium Cesium Cyclotriphosphate Dihydrate

Soufiane Zerraf, Mustafa Belhabra,
Aziz Kheireddine, Malika Tridane,
Hicham Moutaabbid, Mohammed Moutaabbid and
Said Belaaouad

Additional information is available at the end of the chapter

http://dx.doi.org/10.5772/intechopen.81118

Abstract

Chemical preparation, crystal structure, thermal behavior, and IR studies are reported for the barium cesium cyclotriphosphate dihydrate $BaCsP_3O_9.2H_2O$ and its anhydrous form $BaCs_4(PO_3)_6$. $BaCsP_3O_9.2H_2O$, isotypic to $BaTlP_3O_9.2H_2O$ and $BaNH_4P_3O_9.2H_2O$, is monoclinic P21/n with the following unit cell dimensions: a = 7.6992(2)Å, b = 12.3237(3)Å, c = 11.8023(3)Å, α = 90 (2)°, β = 101.18(5)°, γ = 90. (3)°, and Z = 4. The total dehydration of $BaCsP_3O_9.2H_2O$ is between 100°C and 580°C. The IR absorption spectroscopy spectrum for the crystal confirms that most of the vibrational modes are comparable to similar cyclotriphosphates and to the calculated frequencies. The thermal properties reveal that the compound is stable until 90°C.

Keywords: barium cesium cyclotriphosphate, crystal structure, vibrational study

1. Introduction

During a systematic investigation of cyclophosphates, types $BaCsP_3O_9.xH_2O$, $BaCs_4(P_3O_9)_2.xH_2O$, $BaCs_2P_4O_{12}.2H_2O$, $Ba_3Cs_2(P_4O_{12})_2.2H_2O$, and $BaCsP_3O_9.2H_2O$ were obtained. Barium and cesium cyclotriphosphate dihydrate, $BaCsP_3O_9.2H_2O$, was prepared for the first time by using Boulle's process [1] by Masse and Averbuch-Pouchot [2], who described it as a monohydrate. The literature provides $BaCsP_3O_9.2H_2O$ crystallizing in the monoclinic system, space group *P2₁/n*, Z = 4 with the following unit cell parameters, *a = 7.6992(2) Å*, *b = 12.3237(3) Å*, *c* = *11.8023 (3)* Å, and β = *101.181 (5)*° with a brief report of the structural refinement based on single-crystal XRD data. In the present work, we report the chemical preparation, crystalline structure, thermogravimetric analysis, and infrared study of this crystal barium and cesium cyclotriphosphate dihydrate, $BaCsP_3O_9.2H_2O$, in order to have maximum information about structure and reactivity of the solids.

2. Experimental parameters

2.1. Chemical preparation

Single crystals of $BaCsP_3O_9.2H_2O$ were prepared by slowly adding dilute cyclotriphosphoric acid, $H_3P_3O_9$, to an aqueous solution of barium carbonate, $BaCO_3$, and cesium carbonate, Cs_2CO_3, with a stoichiometric ratio of Ba-Cs = 1:1, according to the following chemical reaction:

$$H_3P_3O_9 + BaCO_3 + 1/2Cs_2CO_3 \longrightarrow BaCsP_3O_9.2H_2O + 3/2CO_2$$

The solution was then slowly evaporated at room temperature for 45 days until single crystals of $BaCsP_3O_9.2H_2O$ were obtained. The cyclotriphosphoric acid, $H_3P_3O_9$, used in this reaction was prepared from an aqueous solution of $Na_3P_3O_9$ passed through an ion-exchange resin "Amberlite IR120" [3]. $Na_3P_3O_9$ was obtained by thermal treatment of sodium dihydrogen monophosphate, NaH_2PO_4, at 530°C for 5 h in the air, according to the following chemical reaction [4]:

$$3NaH_2PO_4 \longrightarrow Na_3P_3O_9 + 3H_2O$$

2.2. XRD, crystal data, intensity data collection, and structure

A single-crystal X-ray structure determination of $BaCsP_3O_9.2H_2O$ was performed by using an Oxford Xcalibur S diffractometer at 293 K.

The structure was solved by direct methods using SHELXS [5] implemented in the Olex2 program [6]. The refinement was then carried out with SHELXL by full-matrix least squares minimization and difference Fourier methods. All non-hydrogen atoms were refined with anisotropic displacement parameters. Hydrogen atoms were generated in idealized positions, riding on the carrier atoms, with isotropic thermal parameters.

The final R1 value is 0.0401 for 1782 reflections with $I > 2\sigma(I)$, and full X-crystal data is presented in **Table 1**. The main geometrical features, bond distances, and angles are reported in **Table 6**.

2.3. Fourier transform infrared spectroscopy (FTIR)

A Nicolet Magna IR 560 spectrometer (resolution 1 cm^{-1}, 200 scans) and an OMNIC software were used to characterize the stretching and bending bands between 400 and 4000 cm^{-1}.

http://dx.doi.org/10.5772/intechopen.81118

Compound	2
Empirical formula	BaCs $H_4O_{11}P_3$
Formula weight	543.20 g.mol^{-1}
Crystal system/space group	Monoclinic/*P 2$_1$/n*
a/Å	7.6992(2) Å
b/Å	12.3237(3) Å
c/Å	11.8023(3) Å
α/°	90°
β/°	101.18(3)°
γ/°	90°
V/Å^3	1098.57(5)Å^3
Z	4
D_{calc} (g/cm^3)	3.284 g/cm^3
μ (mm^{-1})	7.362
Crystal size (mm)	0.3296 × 0.1602 × 0.0957 mm^3
Color/shape	Colorless/prism
Temp (K)	293(2)K
Theta range for collection	3.50°/27.59°
Reflections collected	9176
Independent reflections	2448
Data/restraints/parameters	2448/0/147
Goodness of fit on F^2	1.113
Final R indices [I > 2σ(I)]	R_1 = 0.0285, wR_2 = 0.0611
R indices (all data)	R_1 = 0.0329, wR_2 = 0.0638
Largest difference peak/hole	0.78/−1.40 Å^{-3}

Table 1. Crystal data and experimental parameters for the X-ray intensity data collection for $BaCsP_3O_9.2H_2O$.

3. Results and discussion

3.1. Structural analysis

The final atomic positions and anisotropic thermal parameters for the non-hydrogen atoms in the $BaCsP_3O_9.2H_2O$ structure are given in **Tables 2** and **3**, respectively. A projection of the $BaCsP_3O_9.2H_2O$ atomic arrangement along the c axis is given in **Figure 1**. It shows that all the components of the atomic arrangements are located around the two axes in order to form arrays delimiting large channels parallel to the c direction.

Atoms	X	Y	Z	U_{eq}
Ba	0.24946(3)	0.06963(2)	0.37463(2)	0.01486(9)
Cs	1.23531(4)	0.37670(3)	0.60501(3)	0.02500(10)
P(1)	0.49653(15)	0.33939(9)	0.34729(10)	0.0135(2)
P(2)	0.75498(15)	0.17362(10)	0.42595(10)	0.0140(2)
P(3)	0.72984(16)	0.35936(10)	0.57392(11)	0.0185(3)
O(1i)	0.8311(4)	0.2619(3)	0.5238(3)	0.0194(7)
O(2i)	0.6424(4)	0.2510(2)	0.3278(2)	0.0159(7)
O(3i)	0.6022(4)	0.4024(2)	0.4588(3)	0.0178(7)
O(4e)	0.8606(5)	0.4456(3)	0.6136(4)	0.0393(10)
O(5e)	0.6256(5)	0.3191(3)	0.6572(3)	0.0312(9)
O(6e)	0.4740(4)	0.4168(2)	0.2497(3)	0.0212(7)
O(7e)	0.9053(4)	0.1308(3)	0.3805(3)	0.0223(8)
O(8e)	0.6306(4)	0.0994(3)	0.4691(3)	0.0201(7)
O(9e)	0.3428(4)	0.2843(3)	0.3783(3)	0.0205(7)
O(10w)	0.2195(4)	0.1274(3)	0.5953(3)	0.0233(8)
O(11w)	0.5532(5)	0.0975(3)	0.7172(3)	0.0315(9)
H(1)	0.5738	0.0941	0.7926	0.047
H(2)	0.5846	0.1617	0.6363	0.047
H(3)	0.1274	0.1017	0.6242	0.035
H(4)	0.3191	0.0980	0.6366	0.035

i, internal; e, external; w, water.

Table 2. Final atomic coordinates and U-equivalent temperature factors for $BaCsP_3O_9.2H_2O$.

3.2. Barium and cesium arrangement in the structure

The barium atom, located on the twofold axis, is coordinated by two water molecules and six oxygen atoms (**Figure 2**), forming an almost regular dodecahedron. The Ba-O distances spread between 2.298(6) and 2.349(6) Å. Each BaO_8 dodecahedron shares six oxygen atoms with two anionic rings belonging to two phosphoric layers, thus providing the cohesion between these layers (**Figure 2**). BaO_8 dodecahedra do not share any edge or corner and form layers alternating with P_3O_9 ones. The shortest Ba-Ba distance is found to be 4.70731 Å (**Table 4**).

The cesium atom occupies a general position and is coordinated to 10 external oxygen atoms and one water molecule (**Figure 3**). The Cs-O distances spread between 3.0278(2) and 3.5982(9) Å.

The water group, its environment, established by strong hydrogen bonds, is depicted in (**Figure 3**) as an ORTEP representation [7].

http://dx.doi.org/10.5772/intechopen.81118

Atom	U11(s)	U22	U33	U23	U13	U12
Ba	0.01483(15)	0.01281(16)	0.01595(15)	0.00044(10)	0.00056(11)	−0.00029(10)
Cs	0.02411(18)	0.0258(2)	0.02600(18)	0.00406(13)	0.00715(14)	0.00042(13)
P(1)	0.0146(6)	0.0132(6)	0.0122(5)	0.0016(4)	0.0015(5)	0.0020(5)
P(2)	0.0150(6)	0.0130(6)	0.0134(5)	0.0004(5)	0.0016(5)	0.0028(5)
P(3)	0.0171(6)	0.0193(7)	0.0171(6)	−0.0062(5)	−0.0015(5)	0.0022(5)
O(1i)	0.0173(16)	0.0172(17)	0.0207(17)	−0.0054(14)	−0.0035(14)	0.0032(14)
O(2i)	0.0195(17)	0.0166(17)	0.0113(15)	0.0007(13)	0.0023(13)	0.0071(14)
O(3i)	0.0192(17)	0.0140(17)	0.0176(17)	−0.0035(13)	−0.0026(14)	0.0027(14)
O(4e)	0.025(2)	0.027(2)	0.058(3)	−0.0226(19)	−0.0086(19)	−0.0009(17)
O(5e)	0.037(2)	0.037(2)	0.0201(18)	0.0038(16)	0.0079(17)	0.0101(18)
O(6e)	0.0241(18)	0.0194(17)	0.0210(18)	0.0080(14)	0.0067(15)	0.0069(15)
O(7e)	0.0173(17)	0.0245(19)	0.0247(18)	−0.0051(15)	0.0031(15)	0.0064(14)
O(8e)	0.0201(17)	0.0147(17)	0.0253(18)	0.0064(14)	0.0042(15)	0.0009(14)
O(9e)	0.0177(17)	0.0185(18)	0.0255(18)	0.0026(14)	0.0050(15)	−0.0004(14)
O(10w)	0.0190(17)	0.027(2)	0.0236(18)	−0.0012(15)	0.0043(15)	0.0010(15)
O(11w)	0.030(2)	0.034(2)	0.028(2)	−0.0008(17)	−0.0009(18)	−0.0001(18)

i, internal; e, external; w, water.

Table 3. Anisotropic thermal parameters (Å^2) for $BaCsP_3O_9.2H_2O$.

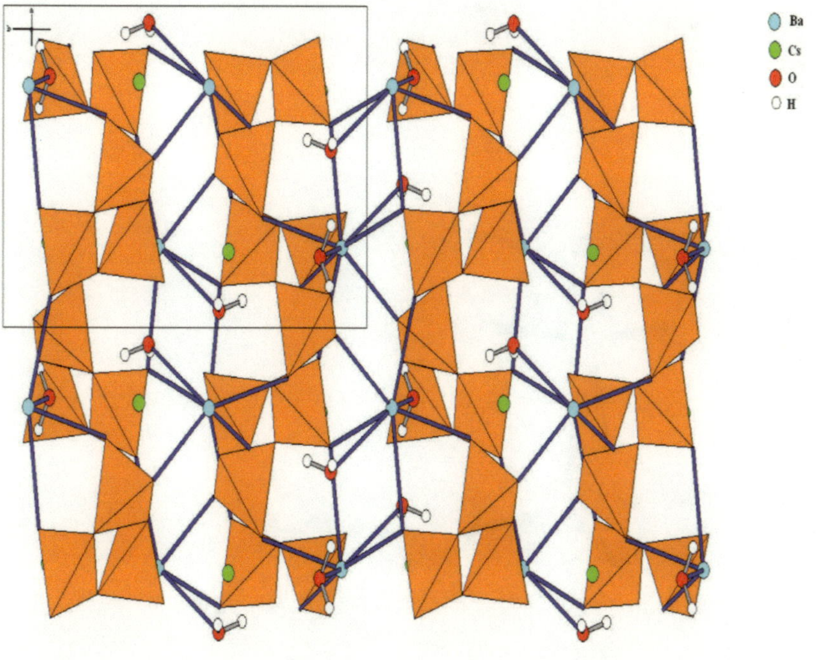

Figure 1. Projection along the c axis of the atomic arrangement in $BaCsP_3O_9.2H_2O$.

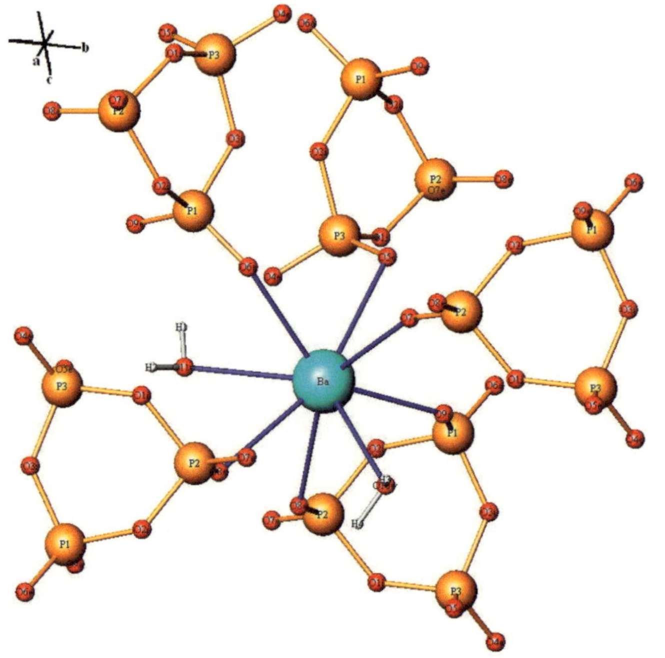

Figure 2. The coordination of the barium atom in $BaCsP_3O_9.2H_2O$.

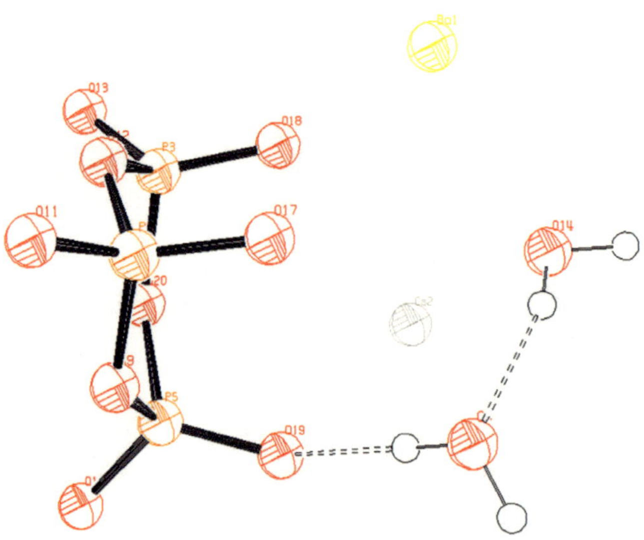

Figure 3. ORTEP representation of $BaCsP_3O_9.2H_2O$ (H-bonds are represented by dashed lines). Thermal ellipsoids are scaled to enclose 50% probability.

Tetrahedron around P(1)				
P(1)	O(2i)	O(3i)	O(6e)	O(9e)
O(2i)	1.6126(5)	100.5(9)	107.8(1)	109.7(7)
O(3i)	2.4804(3)	1.6065(6)	106.9(7)	108.7(6)
O(6e)	2.4983(3)	2.4803(1)	1.4795(3)	120.9(4)
O(9e)	2.5254(9)	2.5046(5)	2.5662(1)	1.4708(8)
Tetrahedron around P(2)				
P(2)	O(1i)	O(2i)	P(2)	O(1i)
O(1i)	1.6120(2)	100.7(7)	107.4(8)	109.7(3)
O(2i)	2.4853(7)	1.6164(2)	107.3(7)	108.6(1)
O(7e)	2.4843(1)	2.4879(6)	1.4650(7)	120.7(8)
O(8e)	2.5346(1)	2.5175(5)	2.5637(8)	1.4838(6)
Tetrahedron around P(3)				
P(3)	(O1i)	(O3i)	(O4e)	(O5e)
O(1i)	1.6058(7)	101.4(4)	107.8(7)	111.2(4)
O(3i)	2.4836(6)	1.6045(7)	107.4(7)	110.5(9)
O(4e)	2.4914(1)	2.48390	1.4762(3)	117.3(4)
O(5e)	2.5386(2)	2.5308(4)	2.5158(1)	1.4705(7)
P(1)–P(2)	2.8773(2)	P(2)–O(1i)–P(3)	129.4(1)	
P(1)–P(3)	2.9289(6)	P(1)–O(2i)–P(3)	131.6(6)	
P(2)–P(3)	2.9081(6)	P(1)–O(3i)–P(2)	125.8(9)	
P(2)–P(1)–P(3)	60.2(1)			
P(1)–P(2)–P(3)	60.7(2)			
P(1)–P(3)–P(2)	59.1(6)			

Table 4. Main interatomic distances (A°) and bond angles (°) in the P_3O_9 ring [8].

3.3. Characterization by infrared spectroscopy

Crystals were ground in a mortar with dry KBr powder in a ratio of 2:200 and pelleted in a press ($8*10^3$ kg, 30 s). Then, they were stored at 95°C for 1 d to dry before use.

The IR spectrum of $BaCsP_3O_9.2H_2O$ illustrated in **Figure 4** reveals the presence of three bands due to water molecules in the domain 4000–1600 cm^{-1}. This confirms the existence of nonequivalent positions of water molecules in the $BaCsP_3O_9.2H_2O$ atomic arrangement: 3449 cm^{-1} attributed to O-H valence vibration, around 3270 cm^{-1} to hydrogen bonds and 1637 cm^{-1} to δHOH deformation. The valence vibration bands related to the P_3O_9 cycles are expected in the domain 1400–650 cm^{-1}, as well as possible bands due to interactions between P_3O_9 cycles and water molecules and also of water vibration modes.

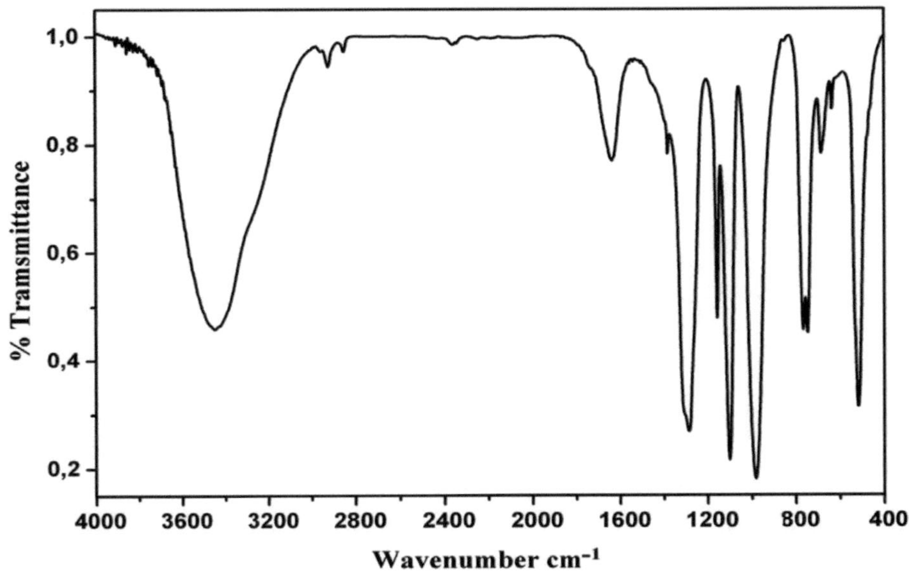

Figure 4. FTIR spectrum of $BaCsP_3O_9.2H_2O$ crystal.

The vibration modes of the phosphate anions usually occur in the 1400–650 cm^{-1} area. The two IR bands observed at 1384 and 1286 cm^{-1} can be attributed to the vas (PO_2) stretching vibration (**Table 5**). The shouldered band at 1157 cm^{-1} and the doublet observed at 1100 and

ν (cm^{-1})	Vibration
3449	ν OH
1637	
	δ HOH
1637	
1384	ν_{as} OPO^-
1286	
1157	ν_s OPO^-
1100	
983	ν_{as} POP
767	ν_s POP
747	
685	
	δ OPO^-
637	+
519	ρ OPO^-

Table 5. Frequencies (cm^{-1}) of IR absorption bands for $BaCsP_2O_9.2H_2O$.

983 cm^{-1} can be assigned to νs(PO_2) and νas(POP), respectively. The most characteristic feature of the P_3O_9 ring anions is the occurrence of a strong intensity band near 767 cm^{-1} in addition to 747 cm^{-1} due to the νs(POP) stretching vibration. The weak peak appearing at 685 cm^{-1} can be assigned to νs(POP) [9]. The broad bands observed at 519 cm^{-1} and the weak peak at 637 cm^{-1} can be due to the deformation vibrations of the anionic group.

In the spectral domain 650–400 cm^{-1}, the spectrum of $BaCsP_3O_9.2H_2O$ (**Figure 4**) shows bending vibration band characteristic of phosphates with ring anions.

4. Vibrational study

The percentage of participation of each group was determined (**Table 6**). The geometrical parameters of the P3O9 3-ring with D3h symmetry, optimized by the MNDO [10] programs, are comparable with those obtained, by X-ray diffraction for the compounds with known structures.

$^{31}P_3{}^{16}O_9{}^{3-}$	$^{31}P_3{}^{18}Oi_3{}^{16}Oe_6{}^{3-}$		$^{33}P_3{}^{16}O_9{}^{3-}$		$^{31}P_3{}^{16}Oi_3{}^{18}Oe_6{}^{3-}$		% of participation
ν(cm^{-1})	ν (cm^{-1})	Δν (cm^{-1})	ν (cm^{-1})	Δν (cm^{-1})	ν (cm^{-1})	Δν cm^{-1})	
1287.75	1287.52	0.23	1269.04	18.71	1249.67	38.08	
1271.80	1271.77	0.03	1253.83	17.97	1233.02	38.78	$\nu_{as}PO_2$ (99)
1271.79	1271.76	0.03	1253.81	17.98	1233.01	38.78	$\nu_{as}PO_2$ (100)
1225.00	1179.05	45.95	1215.39	9.61	1223.98	1.02	$\nu_{as}POP$ (98) + υ_sPO_2 (2)
1224.94	1178.99	45.95	1215.23	9.71	1223.92	1.02	υ_sPO_2 (100)
1168.89	1168.79	0.10	1156.02	12.87	1127.56	41.33	
1108.24	1098.42	9.82	1102.00	6.24	1062.75	45.49	$\nu_{as}POP$ (18) + υ_sPO_2 (82)
1108.21	1098.39	9.82	1101.97	6.24	1062.72	45.49	υ_{as} POP (100)
1059.25	1011.03	48.22	1052.97	6.28	1059.01	0.24	
780.69	768.59	12.10	765.37	15.32	776.16	4.53	υ_sPOP (73) + δPO_2 (27)
780.68	768.57	12.11	765.37	15.31	776.14	4.54	
670.86	659.43	11.43	663.10	7.76	660.19	10.67	υ_sPOP (52) +δPO_2 (48)
558.95	536.78	22.17	555.05	3.90	552.77	6.18	γPOP (60) + γ_RPO_2 (40)
511.25	495.99	15.26	509.05	2.20	501.27	9.98	δPOP (δcycle) (78)
436.70	433.13	3.57	432.42	4.28	422.94	13.76	δPOP (21) + δPO_2 (79)
436.68	433.11	3.57	432.41	4.27	422.92	13.76	
420.07	417.52	2.55	413.15	6.92	411.17	8.90	γ_wPO_2 (78)
418.47	406.18	12.29	416.87	1.60	410.01	8.46	γPOP (59) + γ_TPO_2 (41)
418.41	406.12	12.29	416.81	1.60	409.96	8.45	δPO_2 (98)
301.96	301.61	0.35	301.36	0.60	285.89	16.07	
298.71	292.63	6.08	298.22	0.49	289.41	9.30	δPOP(40) + γwPO_2 (60)
298.67	292.59	6.08	298.18	0.49	289.37	9.30	
280.95	279.15	1.80	279.08	1.87	269.77	11.18	γPOP (14) + γ_TPO_2 (86)
280.92	279.11	1.81	279.05	1.87	269.74	11.18	
256.50	253.00	3.50	255.02	1.48	246.50	10.00	δPOP (26) + γw PO_2 (74)
256.49	252.98	3.51	255.01	1.48	246.49	10.00	γ_TPO_2 (100)
214.13	214.13	0.00	214.13	0.00	201.88	12.25	
49.08	48.30	0.78	49.08	0.00	47.01	2.07	γPOP (27) + γ_RPO_2 (73)
35.78	35.11	0.67	35.77	0.01	34.39	1.39	
34.40	33.75	0.65	34.40	0.00	33.00	1.40	γPOP (33) + γ_RPO_2 (67)

Table 6. IR frequencies and displacements (Δν in cm^{-1}) calculated for the P_3O_9 (D_{3h} symmetry).

All the Raman spectra available in the literature of compounds with the $P_3O_9^{3-}$ cycle of C_{3h} symmetry, in $LnP_3O_9.3H_2O$ [11] and $M^{II}M^{I}P_3O_9$ with benitoite structure 4, and cycle of Cs symmetry in $NiRb_4(P_3O_9)_2.6H_2O$ [17, 18], $ZnM^{I}_4(P_3O_9)_2.6H_2O$ (M^I = K, Rb) [12, 13], $M^{II}K_4(P_3O_9)_2.7H_2O$ (M^{II} = Ni,Co), C_1 in $M^{II}(NH_4)_4(P_3O_9)_2.4H_2O$ (M^{II} = Cu, Co, Ni) [14], and $NiNa_4(P_3O_9)_2.6H_2O$ [15] are characterized by three intense bands situated between 1153 and 1180, 640–680, and 297–313 cm^{-1}, which confirm the results of our calculations (**Table 6**). Indeed, the theory predicts on the whole four bands with A'_1 modes for the P_3O_9 ring with D_{3h} symmetry which are situated, according to our results, at 1169 cm^{-1} for νs P-Oe, 671 cm^{-1} for δs P-Oi, 559 cm^{-1} for δsPOiP, and 302 cm^{-1} for δs PO_2. These four frequencies are predicted to be characteristic in any Raman spectrum of a cyclotriphosphate (with cycle of symmetry, C_3, C_2, Cs, or C_1). These four IR fundamental frequencies have a null calculated intensity and are non-observable for D_{3h} or C_{3h} symmetries, and their appearance in any IR spectrum indicates a symmetry lower than C_{3h}.

M. G.			Δν (cm^{-1})			ν (cm^{-1}) in $BaCsP_3O_9.2H_2O$		
D_{3h}νcal (cm^{-1})	I/Imax	Mode (IR, Ra)				Mode (IR, Ra)		Movement
1287.75	55.3	E"(-,+)	0.23	18.71	38.80	→ A(+,+)	1297	$ν_{as}$ PO_2
								$ν_{as}PO_2$
1271.80	0.00		0.03	17.97	38.78	→ A(+,+)	1274	$ν_{as}$ PO_2
1271.79	0.00	E'(+,+)	0.03	17.98	38.78	→ A(+,+)	1269	
							1211	
		A'_1(-,+)						$ν_{as}POP$
1225.00	100		45.95	9.61	1.02	→ A(+,+)		
1224.94	100	E'(+,+)	45.95	9.71	1.02	→ A(+,+)		$ν_{as}$ POP
1168.89	0.00	A'_2(-,-)	0.10	12.87	41.33	→ A(+,+)	1166	$ν_s$ PO_2
								$ν_sPO_2$
1108.24	5.85	E'(+,+)	9.82	6.24	45.49	→ A(+,+)	1117	
1108.21	5.85		9.82	6.24	45.49	→ A(+,+)	1103	$ν_s$ PO_2
		A'_1(-,+)						
1059.25	0.00		48.22	6.28	0.24	→ A(+,+)	977	
							880	vasPOP
								combination
780.69	18.35		12.10	15.32	4.53	→ A(+,+)	760	
780.68	18.35		12.11	15.32	4.54	→ A(+,+)	743	$ν_sPOP$
								$ν_sPOP$
670.86	0.00		11.43	7.76	10.67	→ A(+,+)	670	

Δν(cm^{-1}): effect of the isotopic substitution; Δν(cm^{-1}): difference between the calculated Value of the frequency before and after the substitution; M. G: molecular group.

Table 7. Attribution of the observed valence IR frequencies (cm^{-1}) of the P_3O_9 ring (C_1) in $BaCsP_3O_9.2H_2O$.

This allowed us an attribution of the 30 fundamental frequencies of the cycle D_{3h} on valid theoretical bases including 12 valence vibration frequencies and 18 bending vibration frequencies. The correlation between the D_{3h} group and the site group C_1 shows that the simple normal modes (A'_1, A'_2, A''_1, and A''_2), of the D_{3h} group, are resolved each into the mode A of the C_1 group and the doubly degenerate E′ and E″ modes are resolved into two modes and are active in IR and Raman. The factor group analysis predicts for four cycles of the unit cells of $BaCsP_3O_9.2H_2O$ (C_{2h}), respectively, 24 and 36 valence vibration bands active in IR. But, we observe in the IR spectra of $BaCsP_3O_9.2H_2O$ (C_{2h}) only six or seven bands and one inflection (**Figure 4**). It seems that the vibrational couplings between the P_3O_9 cycles of the unit cell are absent or very weak; thus, we will be able to interpret the IR spectrum, in the range 1400–650 cm^{-1}, of $BaCsP_3O_9.2H_2O$ according to the vibrations of an isolated cycle with local symmetry C_1. The values of the calculated frequencies, for the D_{3h} symmetry, are close to those observed for $BaCsP_3O_9.2H_2O$ (**Table 6**). **Table 7** gives the attribution of the observed valence frequencies, 1400–650 cm^{-1}, of the P_3O_9 ring, with D_{3h} symmetry of $BaCsP_3O_9.2H_2O$.

5. Thermal analysis

The curve corresponding to the TG analyses in an air atmosphere and at a heating rate of 10°C. min^{-1} of BaCsP3O9.2H2O is given in **Figure 5**. The dehydration of the barium cyclotriphosphate and of cesium dihydrate BaCsP3O9.2H2O is carried out in two steps in two temperature ranges from 105 to 180°C and from 180 to 580°C (**Figure 5**). In the thermogravimetric (TG) curve, the first step between 95 and 180°C corresponds to the elimination of 1.14 water molecules; the second step from 180 to 580°C is due to the removal of 0.86 water molecules.

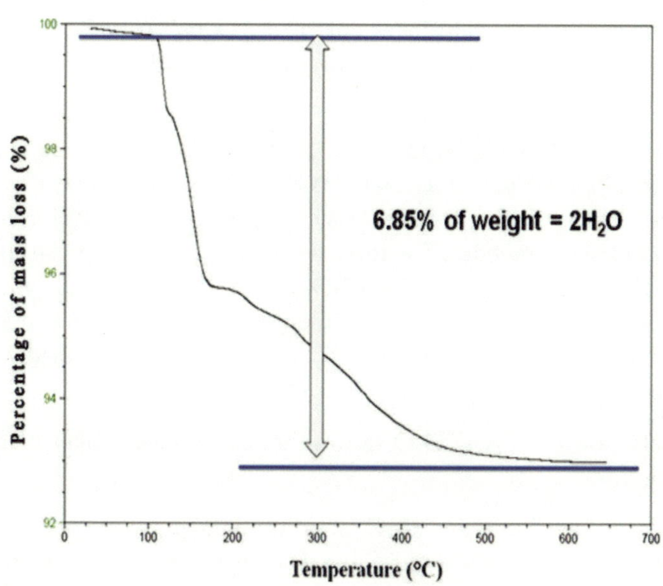

Figure 5. TG curves of $BaCsP_3O_9.2H_2O$ at rising temperature (10°C min^{-1}).

6. Comparison of the thermal behavior of $BaCsP_3O_9.2H_2O$ with $BaNH_4P_3O_9.2H_2O$ and $BaTlP_3O_9.2H_2O$

The thermal behavior of $BaNH_4P_3O_9.2H_2O$ and $BaTlP_3O_9.2H_2O$ [16] was studied (Laboratory of Physical Chemistry of Materials, Ben M'sik faculty of Sciences, Casablanca, Morocco). It would be useful to compare the thermal behavior of $BaCsP_3O_9.2H_2O$ with that of its isotypic compounds $BaNH_4P_3O_9.2H_2O$ and $BaTlP_3O_9.2H_2O$.

The thermal behavior of $BaCsP_3O_9.2H_2O$ is different from that obtained in the case of $BaTlP_3O_9.2H_2O$ [16], which leads to the anhydrous barium and thallium $BaTlP_3O_9$ cyclotriphosphate at 280°C. After amorphous X-ray state, $BaTlP_3O_9$ remains stable till its melting point at 670°C.

$$BaTlP_3O_9.2H_2O \xrightarrow{70-270^\circ C} Amorphous\ X-ray\ Phase + 2H_2O \xrightarrow{300^\circ C} BaTlP_3O_9$$

The total dehydration of $BaCsP_3O_9.2H_2O$, after passing through an amorphous X-ray state, leads to monobarium polyphosphate and tetracerium $BaCs_4(PO_3)_6$ at 500°C [17, 18].

$$BaCsP_3O_9.2H_2O \xrightarrow{100-250^\circ C} Amorphous\ X-ray\ Phase + 2H_2O \xrightarrow{450^\circ C} 1/4\ BaCs_4(PO_3)_6\ crystallized$$
$$+ 3/4\ \left[Ba(PO_3)_2\right]\infty(BaP_2O_6 amorphous\ in\ X-ray\ diffraction).$$

7. Conclusion

The cyclotriphosphate $BaCsP_3O_9.2H_2O$ was obtained as a monocrystal by the resin exchange method. It crystallizes in the monoclinic system, space group $P2_1/n$, Z = 4, and is an isotype of $BaNH_4P_3O_9.2H_2O$ and $BaTlP_3O_9.2H_2O$.

The crystal structure of $BaCsP_3O_9.2H_2O$ was solved from 2448 independent reflections. The final value of the unweighted reliability factor is R = 0.0329. The unit cell of $BaCsP_3O_9.2H_2O$ contains four $P_3O_9{}^{3-}$ rings, each of them consists of three crystallographically independent P (1)O4, P(2)O4, and P(3)O4 tetrahedra. The three tetrahedra have no special characteristics. The P_3O_9 cycle observed in the structure of $BaCsP_3O_9.2H_2O$ has no internal symmetry. The cohesion between the cycles $P_3O_9{}^{3-}$ is ensured via the associated cations Cs^+ and Ba^{2+}. The main geometrical characteristics of the three P(1)O4, P(2)O4, and P(3)O4 tetrahedra of the P_3O_9 cycle are quite similar to those observed generally in cyclotriphosphates.

The thermogram (TG) of $BaCsP_3O_9.2H_2O$ shows that dehydration takes place in two distinct steps between 70 and 560°C.

The total removal of the water at 560°C is accompanied by a total destruction of the $BaCsP_3O_9.2H_2O$ structure, probably leading to a mixture of amorphous oxidesin X-ray diffraction BaO + 3/2 P_2O_5 + ½ Cs_2O. The product resulting from calcinations of $BaCsP_3O_9.2H_2O$ between 300 and 560°C is the long chain polyphosphate $BaCs_4(PO_3)_6$.

Author details

Soufiane Zerraf[1]*, Mustafa Belhabra[1], Aziz Kheireddine[1], Malika Tridane[1,2], Hicham Moutaabbid[3], Mohammed Moutaabbid[1] and Said Belaaouad[1]

*Address all correspondence to: soufiane.zerraf@gmail.com

1 Laboratory of Physical Chemistry of Materials, Ben M'sik Faculty of Sciences, Casablanca, Morocco

2 Regional Center for Education and Training Occupations Casablanca Anfa, Bd Bir Anzarane Casablanca, Morocco

3 Institute of Mineralogy, Physics of Materials, and Cosmochemistry (IMPMC) Sorbonne Universities—UPMC University Paris, Cedex, France

References

[1] Kheireddine A, Tridane M, Belaaouad S. Powder Diffraction. 2012;**27**:32-35. DOI: 10.13171/mjc.2.4.2013.22.05.12

[2] Masse R, Averbuch-Pouchot MT. Materials Research Bulletin. 1977;**12**:13-16. DOI: 10.1016/0025-5408(77)90083-6

[3] Simonot-Grange MH. Journal of Solid State Chemistry. 1983;**46**:76-86. DOI: 10.1016/0022-4596(83)90128-7

[4] Durif A. Crystal Chemistry of Condensed Phosphates. New York: Plenum Press; 1995. ISBN 978-1-4757-9894-4

[5] Sheldrick GM. Acta Crystallographica. 2008;**64**:112-122. DOI: 10.1107/S0108767307043930

[6] Dolomanov OV, Bourhis LJ, Gildea RJ, Howard JA, Puschmann H. Journal of Applied Crystallography. 2009;**42**:339-341. DOI: 10.1107/S0021889808042726

[7] Tarte P, Rulmont A, Sbai K, Simonot-Grange MH. Spectrochimica Acta. 1987;**43A**:337-345. DOI: 10.1016/0584-8539(87)80114-9

[8] Zerraf S, Belhabra M, Kheireddine A, Lamsatfi R, Tridane M, Moutaabbid H, Baptisteb B, Moutaabbid M, Belaaouad S. Phosphorus Sulfur Silicon and the Related Elements. 2017; **192**:1286-1293. DOI: 10.1080/10426507.2017.1333507

[9] Tace EM, Charaf A, Fahim I, Moutaabbid M, Kheïreddine A, Ouaalla F-E, et al. Phosphorus, Sulfur and Silicon and the Related Elements. 2011;**186**:1501-1514. DOI: 10.1080/10426507.2010.520173

[10] Dewar MJS, Thiel W. Journal of the American Chemical Society. 1977;**99**:4899-4902. DOI: 10.1021/ja 00457a004

[11] Sbai K, Abouimrane A, Lahmidi A, El Kababi K, Hliwa M, Vilminot S. Annales de Chimie Science des Materiaux. 1999;**25**(supp. 1):139-143

[12] Abouimrane A. Thèse de Doctorat. Casablanca, Maroc; 2000

[13] Abouimrane A, Sbai K, El Kababi K, Lahmidi A, Atibi A, Vilminot S. Phosphorus, Sulfur and Silicon and the Related Elements. 2001;**179**:69-83

[14] Sbai K, Abouimrane A, El Kababi K, Vilminot S. Journal of Thermal Analysis and Calorimetry. 2002;**68**:109-122. DOI: 10.1023/A:10149288

[15] Sbai K, Atibi A, Charaf A, Radid M, Jouini A. Annales de Chimie Science des Materiaux. 2001;**26**:45-61

[16] Belaaouad S, Sbai KJ. Journal of Physics and Chemistry of Solids. 2003;**64**:981-991. DOI: 10.1016/S0022-3697(02)00461-4

[17] Averbuch-Pouchot MT, Durif A. Acta Crystallographica. 1986;**C42**:928-930. DOI: 10.1107/S0108270186093976

[18] Belhabra M, Kheireddine A, Moutaabbid H, Baptiste B, Lamsatfi R, Fahim I, et al. Research Journal of Recent Sciences. 2015;**6**(Issue 12):7832-7836. ISSN: 0976-3031